Lecture Notes in Mobility

Series Editor

Gereon Meyer ⓘ, *VDI/VDE Innovation + Technik GmbH, Berlin, Germany*

Editorial Board Members

The book series Lecture Notes in Mobility (LNMOB) reports on innovative, peer-reviewed research and developments in intelligent, connected and sustainable transportation systems of the future. It covers technological advances, research, developments and applications, as well as business models, management systems and policy implementation relating to: zero-emission, electric and energy-efficient vehicles; alternative and optimized powertrains; vehicle automation and cooperation; clean, user-centric and on-demand transport systems; shared mobility services and intermodal hubs; energy, data and communication infrastructure for transportation; and micromobility and soft urban modes, among other topics. The series gives a special emphasis to sustainable, seamless and inclusive transformation strategies and covers both traditional and any new transportation modes for passengers and goods. Cutting-edge findings from public research funding programs in Europe, America and Asia do represent an important source of content for this series. PhD thesis of exceptional value may also be considered for publication. Supervised by a scientific advisory board of world-leading scholars and professionals, the Lecture Notes in Mobility are intended to offer an authoritative and comprehensive source of information on the latest transportation technology and mobility trends to an audience of researchers, practitioners, policymakers, and advanced-level students, and a multidisciplinary platform fostering the exchange of ideas and collaboration between the different groups.

Imre Keseru · Annette Randhahn

Editors

Towards User-Centric Transport in Europe 3

Making Digital Mobility Inclusive and Accessible

Editors
Imre Keseru
Vrije Universiteit Brussel
Brussels, Belgium

Annette Randhahn
Mobility, Energy and Future Technologies
VDI/VDE Innovation + Technik
Berlin, Germany

This work was supported by European Commission

ISSN 2196-5544 ISSN 2196-5552 (electronic)
Lecture Notes in Mobility
ISBN 978-3-031-26157-2 ISBN 978-3-031-26155-8 (eBook)
https://doi.org/10.1007/978-3-031-26155-8

This Springer imprint is published by the registered company Springer Nature Switzerland AG
The registered company address is: Gewerbestrasse 11, 6330 Cham, Switzerland

Preface

Mobility is becoming increasingly digitalised. Digitalisation promises many benefits including more convenience for the user in the different stages of the travel process, i.e. providing real-time information, enabling booking, payment and ticketing for a service or vehicle online, or the ability to provide feedback after reaching the destination. There are many benefits to transport operators as well, in terms of cost savings, more efficient use of resources, better management of operations and the provision of more responsive customer service. Digitalisation has been a gradual process in public transport where transition from paper-based ticketing to electronic and mobile ticketing has taken several decades and it is still ongoing. Nevertheless, the digital transition has also enabled the recent emergence or proliferation of novel, 'disruptive' transport services such as free-floating car sharing, shared bicycles and e-scooters, ridesourcing, Mobility as a Service and autonomous shuttles. These services are often exclusively only available through digital channels such as a smartphone application or website. A similar digital revolution has taken place in logistics and especially in online shopping and the related delivery services. In the past decade, digitalisation has enabled the emergence of application-based delivery services such as *UBER Eats* or *Deliveroo* and new ways to collect one's deliveries from smart delivery boxes.

If we look at the other side of the coin, i.e. who can benefit from the advantages of digitalisation, we see that despite the fast proliferation of Internet and mobile services, around 20% of European households still do not have access to broadband Internet and mobile broadband use also shows a high variation across the EU. The necessary infrastructure is usually rolled out in urban areas first, so especially in rural areas, Internet speeds can be below average. Furthermore, new digital mobility services are usually introduced in urban areas, so residents of rural areas are not able to use them. Moreover, 10% of EU residents have never used the Internet and there are many millions without state-of-the-art smartphones and access to a credit card (European Commission, 2022). In addition to technical causes, there are also social or health-related reasons for the digital gap. For example, digital services are rarely designed in such a way that visually impaired people can use them. The interface is also often complicated and designed with multiple levels, e.g. when displaying ticket fares, so older persons or cognitively impaired people become confused. Or it is simply only possible to pay the digital services by credit card, therefore people without one are excluded.

The above data shows that there is an increasing digital gap that may have an impact on who, how and when can access increasingly digitalised mobility services. While the limited access to transport services can lead to transport disadvantage and eventually social exclusion (Schwanen et al., 2015), the increasing digital gap may exacerbate transport disadvantage. Nevertheless, there is still limited evidence of the possible direct link between digital exclusion and transport-related social exclusion (Durand et al., 2022).

Recognising this gap, some recent initiatives have set out to investigate the needs of digitally excluded people, the requirements that the digital mobility system places on them and develop guidelines and tools for policy makers, service and software developers and transport operators to bridge this gap.

Between 2020 and 2022, three projects on inclusive digital transport were funded by the Horizon 2020 programme of the European Union. This book is the initiative of the Inclusive Digital Mobility Solutions (INDIMO) project, a research and innovation action that had the following objectives:

- To improve the understanding of the users' needs towards the digital transport system.
- To improve knowledge about users' requirements in personalised digital transport systems.
- To co-create tools that can help engineers, developers, operators and policy makers to generate an inclusive, universally accessible personalised digital transport system.
- To foster the universal design approach throughout the planning and design process of digital applications and services, both for accessibility and inclusion.
- To influence future policy by feeding project results into European, regional and local policy making.

Several chapters in this book will highlight key methodological achievements and results of the project. In addition, contributions from INDIMO's sister projects Digital Transport in and for Society (DIGNITY)[1] and Transport Innovation for Persons with Disabilities Needs Satisfaction (TRIPS)[2] will highlight additional aspects of the digital mobility gap and the role of collaboration with stakeholders in bridging it. Furthermore, several chapters explain the theoretical foundations and practical application of the capabilities approach and provide further examples of inclusive design written by experts and practitioners.

The book is divided into five parts. In part 1, three chapters examine **the emergence of the new digital ecosystem in transport**. Teresa de la Cruz et al. analyse the current transition that urban mobility systems are undergoing and present an approach for guiding cities towards the implementation and adoption of new digital urban mobility solutions. The second chapter by Annette Randhahn et al. investigates opportunities and risks of automated and autonomous vehicles (AVs) and the benefits that co-creation and universal design can provide related to autonomy and independence of people in vulnerable situations. The third chapter, written by Xavier Sanyer Matias and Lluís Alegre Valls, demonstrates the evolution of the digital mobility ecosystem in the Barcelona Metropolitan Region through an analysis of the planning approach, the emergence of new, digital mobility options like shared e-scooters, and the user-centric digitalization strategy of mobility in the region.

In the second part of the book, we explore the **reasons and evidence for digital exclusion in transport** through both theoretical and practical approaches. The chapter by Lluis Martinez and Imre Keseru reviews the literature on transport disadvantages, digital exclusion in the context of shared transport. They propose a comprehensive approach to the study of digital shared mobility services by incorporating the digital

---

[1] Project website: https://www.dignity-project.eu
[2] Project website: https://trips-project.eu/

divide into the capabilities approach. In the chapter written by Suzanne Hiemstra-van Mastrigt et al., the authors build further on a needs-based approach. They categorise the different groups of non-digital travellers and create five need-based personas in the context of demand-responsive transport (DRT) services, where the digitalization of the booking process is quite advanced. Based on this, they formulate user requirements and design recommendations for mobility services, and for DRT services specifically. The chapter of Floridea Di Ciommo et al. demonstrates empirical insights from the INDIMO project building on the capabilities approach. They propose the integration of the analysis of capabilities, limitations and requirements of users and non-users into the assessment of the needs of persons in vulnerable situation in the digital mobility context. They also identify three key needs of vulnerable groups towards digital transport solutions.

In part 3, we investigate the **role of participatory planning and design methods in making digital mobility more inclusive and accessible**. The first chapter by Kathryn Bulanowski et al. demonstrates practical methods on how to involve people in vulnerable situation in the co-design of new services through interviews and surveys. The second chapter by Floridea Di Ciommo and colleagues gives us insights into the co-creation process applied in the INDIMO project through the communities of practice approach. Communities of practice bring together different stakeholders who share a common interest and want to take action together. In the concluding chapter in part 3, Michael Abraham and Carolin Schröder present a space transformation experiment that was conducted in Berlin to promote modal shift to alternative mobility services including new, digital mobility options such as rental bicycles, e-mopeds, e-scooters and free-floating car sharing that are available at a mobility hub. An impact and process evaluation provides insights into the usefulness, acceptance and spatial impact of a mobility station integrating several digital mobility solutions.

Part 4 focuses on **user-centric design approaches that can ensure that the needs of users are taken into account when developing digital mobility solutions**. The chapter written by Jørgen Aarhaug introduces universal design, an approach to design products and services to be usable by all, as much as possible, without the need for adaptation or specialised design (Mace, 1998). In the next chapter, Vasconcelos et al., propose a co-design approach for inclusive transport services. Participatory Design Research builds on the assumption that people are the best judges of their own needs and requirements. They offer some copying mechanisms to address the challenges of co-creation and co-design. While the benefits of user-centric research approaches are numerous, the authors also highlight the challenges that researchers and participants might face in such a process. The chapter of Giorgi et al. proposes a user-centric approach and four-step methodology to the development of universally accessible pictograms in digital mobility interfaces. It relies on quick and simple exercises that involve potential users of the icons to evaluate how understandable they are. The research led to one of the main outputs of the INDIMO project, the Universal Interface Language for Digital Mobility Services, which includes an icon catalogue.

The chapters in the final part of the book propose and discuss **various methods, tools and insights of how policy makers can be supported to make digital mobility services more inclusive and accessible**. In the chapter written by Julia Hansel and Antonia Graf, they apply qualitative content analysis of policy documents and ethnographically oriented observation to identify which knowledge, qualifications and resources users as subjects of new technologies should possess to be able use them. The next chapter by Nina Nesterova and colleagues introduces the approach that the DIGNITY project developed to identify digital mobility gaps, which includes a self-assessment framework that can be used by public authorities to identify potential gaps in the development of local digital transport ecosystems. The framework was demonstrated in Barcelona, Tilburg, Flanders and Ancona. The next chapter by Hannes Delaere et al. proposes a new evaluation tool for inclusivity and accessibility of digital mobility and delivery services developed in the INDIMO project. The chapter explains the methodology of the tool development and its testing. It also outlines how the tool can help policy makers to assess the inclusivity of a new or existing service. This paper is complementary to the paper of Nesterova et al. While the former offers a tool to policy makers in the initial phase of identifying the digital gap, the Service Evaluation Tool developed in the INDIMO project offers a self-evolution tool to assess actual digital services and provides recommendations on how to improve them. In the final chapter, Alexandra Pinto et al. investigate the widespread introduction of digital applications that was triggered by a major German government funding programme aiming at enhancing digitalisation in German municipal transport systems. They address in particular the challenges vulnerable groups face in this digital transformation and identify the potential of digital tools to contribute to improved inclusiveness.

We hope that the book gives a wide panorama of the research and implementation issues that surround the digitalisation of mobility and logistics and the contributions will open new avenues for research and innovation in future.

Imre Keseru
Annette Randhahn

References

Durand, A., Zijlstra, T., Van Oort, N., Hoogendoorn-Lanser, S., Hoogendoorn, S.: Access denied? Digital inequality in transport services. Trans. Rev. **42**(1), 32–57 (2022). https://doi.org/10.1080/01441647.2021.1923584

European Commission. Digital Economy and Society Index (DESI) 2022 Digital infrastructures. European Commission (2022). https://ec.europa.eu/newsroom/dae/redirection/document/88766

Mace, R. L.: Universal Design in Housing. Assistive Technol, **10**(1), 21–28 (1998). https://doi.org/10.1080/10400435.1998.10131957

Schwanen, T., Lucas, K., Akyelken, N., Cisternas Solsona, D., Carrasco, J.-A., Neutens, T.: Rethinking the links between social exclusion and transport disadvantage through the lens of social capital. Trans. Res. Part A: Policy Pract. **74**, 123–135 (2015). https://doi.org/10.1016/j.tra.2015.02.012

Contents

The Emergence of a New Digital Mobility Ecosystem

Urban Mobility Transition Driven by New Digital Technologies

Teresa de la Cruz[✉], Beatriz Royo, and Carolina Ciprés

Zaragoza Logistics Center, Zaragoza, Spain
{mdelacruz,broyo,ccipres}@zlc.edu.es

Abstract. The urban mobility landscape for both, freight and passengers, is in transition. During the last decade new business models, enabled by digital technologies, are blooming. However, sometimes the new mobility solutions do not fit with local regulations, their impacts are unclear, and legislative issues are hindering the economic niche exploitation and their implementation. This chapter describes the current transition that urban mobility systems are undergoing and presents an approach for guiding cities towards the implementation and adoption of new digital urban mobility solutions. This approach, based on the Horizon 2020 SPROUT project, consists of assessing the impacts and feasibility of the new mobility solutions, identifying areas where policy intervention to enable the implementation would be required and co-creating those specific policies with all the urban mobility stakeholders. This is complemented by an implementation feasibility and user acceptance analysis.

1 Introduction

The digital revolution started in the middle of the last century, integrating digital technologies into everyday life. This megatrend is known as digitalization and in the last decade has re-shaped the urban mobility landscape. We have seen worldwide emerging in our cities new business models enabled by digital technologies: online marketplaces for carpooling or carsharing, bike sharing systems, etc. But not only passenger transport has experienced the digital revolution. There are also new digital business models for hyper-local logistics, crowd-shipping services for last mile delivery and cargo-hitching.

Very often the new mobility services face regulatory challenges. A good example of this are the free-floating scooters that flooded cities operating under different companies only to face later bans and restrictions. Legislative and governmental issues are thus affecting the implementation of new business models [1], sometimes also hindering economic opportunities.

The creation of new job profiles also impacts on the quality of the labour market and its regulations. Specifically, there is a heated social debate in Europe regarding the platform economy[1] and its link to precarious working conditions [2]. Countries such

---

[1] Platform work is non-standard work facilitated by online platforms which use digital technologies to 'intermediate' between individual suppliers (platform workers) and buyers of labour [2].

I. Keseru and A. Randhahn (Eds.): *Towards User-Centric Transport in Europe 3*, LNMOB, pp. 3–21, 2023.
https://doi.org/10.1007/978-3-031-26155-8_1

as The Netherlands, Germany, France, Spain, Italy or the UK have already approved national laws for regulating the employment status of workers in the platform economy.

Concerning the social dimension, there is an additional debate related to the digital divide between digital savvy users and vulnerable users. This concerns not only mobility but also other universal services such as finance, healthcare or education. The group of users in risk of exclusion includes but is not limited to those without access to meaningful connectivity (rural areas, low-level economies), with lack of digital skills (older people), with language issues (migrants), experiencing cultural barriers (ethnic minorities) or having a negative security perception (women). The European Union has recently introduced the concept of industry 5.0 with the aim to promote the inclusivity in the industry strategic plans when embracing digital and green technologies[2]. It claims for *"social innovation to enhance prosperity and foster good quality jobs alongside measures to support education and skill training to enable workers to adapt to a shifting job market. This includes access to technology to avoid digital gaps in regions with less industrial development and the creation of employment and opportunity with a focus on ensuring economic security and social justice at the same time. Equal access to education and healthcare as well as safeguarding social mobility are fundamental roles of governments (at least in Europe) and vital prerequisites for revolutionising industry and making it people- and planet-proof."*

Generally speaking, the impacts of emerging transport solutions are unclear and therefore not being addressed or are inadequately addressed by the current urban policy instruments. Consequently, the traditional regulatory role of cities is not adequate anymore.

Take the example of the e-scooters. By banning new a mobility solution, cities not only miss the train towards sustainability, but also evidence difficulties in managing and adopting innovation. Uptake of mobility innovations requires more flexible and adaptable local administrations. In contrast, local authorities sometimes lack the capacity to understand the technological and financial implications of innovations and integrate fact-based evidence into decision-making. Therefore, there is a need to provide urban planners and policymakers with tools to navigate urban mobility policy through the transition.

This chapter addresses an approach for guiding cities towards the implementation and adoption of new digital urban mobility solutions.

Section 2 illustrates the transition that urban mobility is currently experiencing and identifies its main driving elements. Section 3 aims to provide some enablers or facilitators for the transition to be innovative and sustainable. The aim of Sect. 4 is to provide cities with guidance to set the policy response and ensure successful adoption when introducing new mobility solutions. Finally, Sect. 5 draws the main conclusions of the chapter and proposes a path ahead.

2 Urban Mobility Transition

The evolution of urban mobility is based on the interplay between different factors. On the 'demand' side contributing factors include varying demographic patterns linked to

---

[2] https://ec.europa.eu/info/publications/industry-50.

economic growth and societal changes, resulting in new patterns of consumption. On the supply side, changes in transport infrastructure provision are often associated with advances in technology. Transport policy plays a major role in this transition, by funding major transport investments and through the introduction of a broad range of physical, regulatory and pricing measures. Such measures have also evolved over time and have been introduced in response to a changing set of perceived concerns, policy objectives and policy priorities [3].

Currently, unprecedented transformations are going on in the realm of urban mobility of passengers and goods. Socio-economic changes and technological advances have resulted in a state in which transport supply and demand are constantly shifting.

The latest financial crisis created a supply-side push from people seeking work opportunities and a demand-side pull from consumers seeking cheaper alternative transportation services. This accelerated the emergence of the collaborative economy, unlocking the value of existing resources and avoiding the need for additional capital expenditure.

On the other hand, the development of digital technologies and widespread Internet access have created new opportunities to make the existing transportation network more efficient and tailored to the need of different users. The concept Mobility as a Service (MaaS)[3] has become popular, fuelled by countless of innovative new mobility services such as carpooling and ridesharing, micro mobility and carsharing systems as well as on-demand "pop-up" bus services. Furthermore, the trend is motivated by the anticipation of self-driving cars, which are expected to change car ownership.

Likewise, consumers' habits are also shifting towards on-demand solutions able to satisfy their needs for faster delivery. Digital technologies contribute to make same day deliveries a reality as well. Similar to the MaaS concept, Logistics as a service (LaaS) is gaining momentum due to the rise of the on-demand economy in urban logistics. Sustainable last-mile logistics offerings are also appearing such as e-vehicles, crowd shipping, crowdsourcing, physical internet,

The COVID-19 pandemic has shown practically that future mobility systems could be very different. It is not yet entirely clear how the "new normal" will look like, but the pandemic has highlighted the importance of issues such as health, hygiene, the environment and home life, as well as speed, convenience, accessibility, inclusivity and consumption. Soft (and also healthier and greener) modes of passengers' mobility such as walking and cycling have become more attractive that, if supported by the reallocation of space, could permanently change travel behaviour.

Many projects funded by the European Union implement pilots and Urban Living Labs to develop, test, and validate new mobility solutions and unleash their potential (Fig. 1). Living Labs are generally defined as '*user-centred, open innovation ecosystems based on systematic user co-creation approach, integrating research and innovation processes in real life communities and settings*'[4]. Living Labs operate as intermediaries among administrations in urban and peri-urban areas, public and private operators, start-ups, third sector and research organisations as well as citizens for joint value co-creation, rapid prototyping or validation to scale up and speed up innovation and businesses[5]. So

---

[3] https://maas-alliance.eu/.

[4] https://enoll.org/about-us/.

[5] https://digital-strategy.ec.europa.eu/en/news/living-labs-and-open-innovation.

far, most of the developments have focused on decarbonizing and digitalizing urban mobility.

Fig. 1. Run session trial with two paired electric autonomous pods (NEXT system) for cargo hitching in Padua (SPROUT project)

City administrations and authorities can influence this transition by developing policies that regulate the new mobility ecosystem and enable other actors. Examples of framing actions for authorities are urban space allocation and regulation, infrastructure & data regulation and enforcement of regulations. On the other hand, authorities can enable other actors through governance, infrastructure (physical and digital), mobility demand incentives and marketing campaigns, and collaborative platforms and innovation [4].

The abovementioned actual and foreseen changes in urban mobility are motivated by different drivers. Figure 2 shows the urban mobility catalogue of transition drivers identified by the Sustainable Policy Response to Urban Mobility Transition (SPROUT)[6] project under 6 categories following the PESTEL approach (Social, Technological, Economic, Environmental, Political, and Legal) including also those concerning inclusivity. It was found that the considered importance of drivers differs significantly from city to city depending on the specific characteristics and peculiarities. Considering the larger driver categories, environmental and technological were considered the most important, but when considering individual, 'political agenda' (category of political drivers) and 'urban structure' (category of social drivers) were considered the most important [5].

A Eurobarometer survey showed in 2013 that there was an increasing 'urban mobility gap' between Europe's few advanced cities and the majority trailing behind[7]. This gap has not been closed yet showing that there is a need for reinforcing the support to European cities for addressing urban mobility challenges. There is still no clear trend towards

---

[6] https://sprout-civitas.eu/.

[7] Communication from the Commission to the European Parliament, the Council, the European Economic and Social Committee and the Committee of the Regions Together towards competitive and resource-efficient urban mobility /* COM/2013/0913 final */.

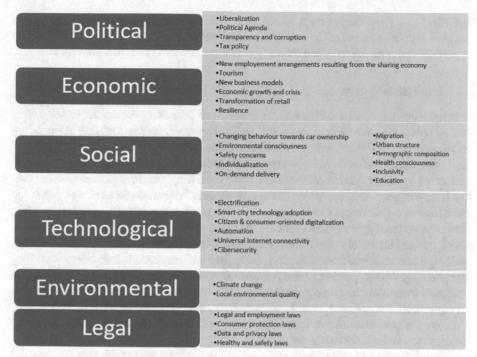

Fig. 2. PESTEL categorisation of urban mobility drivers (adaptation of SPROUT project [5])

more sustainable modes of transport. Overall, there has been no significant reduction in private car use, there are still many cities exceeding EU minimum air quality standards, greenhouse gas emissions due to road transport have been steadily increasing over time and travel by public transport usually takes significantly more time than by private car [6].

The approach to urban mobility is required also ensure that Europe's urban areas develop along a more sustainable path and that EU goals for a competitive and resource-efficient European area are met [7]. At the same time, the transition toward digital mobility services must ensure that the digital divide is not growing but shrinking.

3 Enablers for a Good Transition in Urban Mobility

Evidence-based policymaking has two goals: to use what we already know from programme evaluation to make policy decisions and to build more knowledge to better inform future decisions [8].

There is an increasing demand for the use of evidence to fight against a post-fact/fake news world, and to design more effective policies and better align resources. However, very often, in reality, evidence competes with values, feelings, and emotions (of politicians and citizens), resulting in good evidence as only one element in political decision making [9]. Some enablers to ensure a good transition towards a more innovative and sustainable urban mobility are: an existing innovation ecosystem, good data availability and analytical skills, the engagement of citizens, vulnerable users, and other stakeholders, political support, and access to the right funding solutions. Capacity of local authorities to prepare and implement urban mobility measures and strategies is a prerequisite.

3.1 Innovation Ecosystem

Urban innovation can be defined as *'A break from common practice to develop long lasting transformations in communities, neighbourhoods, and cities'* [10]. Depending on the approach followed, innovation can be incremental, breakthrough or radical.

Cities, rather than national governments, are more likely to lead change and innovation in the transport system, as they have more regulatory freedom to deal with innovative transport providers, are aware of the city-specific innovation aspects, and can at the same time stimulate urban mobility innovation and ensure the delivery of social benefits. However, national governments provide the legislation within which urban transport is developed and the regulatory framework where it operates; they determine the decision-making framework within which cities formulate and implement transport plans; they allocate a significant portion of the finance for urban transport, specify how it may be used, and determine the other ways in which cities can seek funding. Generally, regulatory approaches at national level are different, focusing on issues such as market access, employment and taxation, while leaving the equally important policy challenges at the local/urban level widely unaddressed [11].

To unlock the benefits of new mobility solutions, legal and regulatory frameworks in cities and administrations must be more flexible and adaptable [12]. Across Europe, a common challenge for projects implementing new solutions is working within legal frameworks that support traditional planning methods and are not adapted to innovations in technology and urban planning [13]. Laws and regulations restricting the deployment of autonomous vehicles, electric vehicles, new market solutions, data management, building codes, and even parking, can make it difficult to implement new projects in urban environments. Even if there is political commitment, it can take extensive periods of time to adjust the legal and regulatory frameworks to implement projects [14].

Innovation deployment depends on the right conditions being in place - for instance Living Labs and large-scale demonstrations that help raise political support for sustainable mobility, and to secure investments in sustainable mobility measures. This needs to be complemented by lean procedures to facilitate the approval and deployment of urban mobility innovations, granting permits and exceptions through regulatory sandboxes where relevant.

Aptitude and readiness of cities towards urban mobility innovation varies among cities. It depends on different factors such as inter-departmental coordination in public administration, sustainability awareness of the citizens, skilled workforce on data analytics, knowledge transfer activities, participatory practices with stakeholders or open data availability [15].

3.2 Data and Analytical Skills

When cities imagine the future of mobility, frequently one of the most plausible scenarios consists of optimized and integrated mobility systems and tailored offers to citizens' needs [16]. Comprehensive mobility data analysis is a pre-requisite for the realization of such a vision [13].

Transition requires a modern toolbox of solutions for collecting, managing, and sharing data. Understanding how people and goods move through the city is crucial to help implement the right solutions in the right places for the right target groups. Digitalization and new survey methods sustained by a technology ecosystem that supports data collection with the proper compiling, managing, understanding, and analysis of data also allow to build a more accurate picture of how specific social or demographic groups travel. The whole ecosystem of technology for data collection needs to be considered to make the most of the data available and create better visibility of movements throughout the city. More accurate data also will later allow to better measure the effects of implemented solutions.

Implementing an effective urban mobility policy becomes even more challenging in the case of urban freight, as accessing data is problematic. Little ongoing public data collection about urban freight operations occurs [17]; to a large extent due to the commercially sensitive nature of freight data and the required involvement of a large number of economic actors in a fragmented industry. In addition, there are no standards in Europe that would unify the way of gathering the data collected. As a consequence, urban freight policy is too often based on insufficiently detailed analysis and repetition of regulatory initiatives regardless of local characteristics and dynamics. As opposed to urban passenger transport, there is a lack of national or regional bodies dealing with city logistics [18].

Partnerships on data collection and management across the knowledge triangle (Research, Education and Innovation) and cities are instrumental to the provision of reliable and seamless mobility services, as well as data sharing agreements between private actors operating mobility solutions and public administration.

Given the increasing availability of open data sets, and real-time information from sensors and Internet of Things (IoT) and other applications, cities are more and more operating in a smarter way. Current lack of data could be reversed to an excess of data situation, leading to the 'data paradox' where there is too much generated data but too little of the right data. This will bring the urgent need for building local capacity, providing city managers with tools to help make sense of these data flows.

3.3 The Engagement of Citizens and Other Stakeholders

Acceptance of a policy by citizens and other stakeholders can be enhanced by consultation. This evidence stresses the need for the stakeholders' engagement as a strategic factor of any decision-making process [19].

Stakeholders' involvement can be represented as a pyramid. At the top of there is participation either in decision-making, defining objectives or project elaboration (Fig. 3).

Governance is one of the key aspects of sustainable urban development, as good governance arrangements can contribute to more transparent, inclusive, responsive and effective decision-making. The three central components of the sustainable urban development process are [21]:

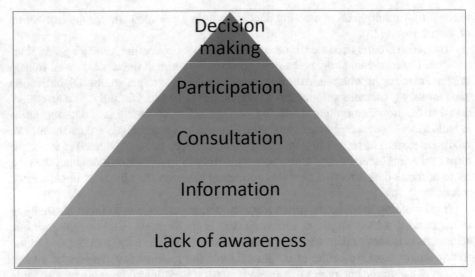

Fig. 3. Community involvement in urban projects pyramid [20]

- multi-level governance, referring to the coordination and alignment of actions (interventions) between different levels of government;
- a multi-stakeholder approach, referring to the inclusion of all relevant actors throughout the whole policy cycle;
- a bottom-up and participatory approach, referring to the use of community-led initiatives to encourage local actors' involvement and response.

The concept of Sustainable Urban Mobility Plans (SUMP), which are promoted by the European Commission (EC), establishes the principle that the society should be involved from the very beginning of the planning process. SUMP is a strategic and integrated approach for dealing with the complexity of urban transport. Its main objective is to improve accessibility and quality of life by achieving a shift towards sustainable mobility. SUMP advocates for fact-based decision making guided by a long-term vision for sustainable mobility [21]. The EC introduced the concept of SUMP with its 2013 Urban Mobility Package [22] in an attempt to address urban mobility challenges and deficient planning practices on the local level. Since then, SUMPs have been established as a concept in many of Europe's larger and medium-sized cities, and capacity in cities has been improved over the years. Yet, for SUMPs to be successful, they need to be the output of a process that involves many stakeholders and requires sufficient resources[8]. Moreover, both passengers and freight transport ecosystem, including stakeholders should be considered from the early stages of the SUMP development to increase its impact and ensure that issues related to emissions and congestion, safety, cost-effectiveness and economic development are fully addressed [23].

---

[8] https://www.polisnetwork.eu/document/joint-stakeholder-statement-on-eu-urban-mobility-fra mework/.

Stakeholders´ involvement and citizens´ participation practices in transport planning differ between European countries and between cities. Several countries have formal, mandatory consultation procedures for mid- and large-scale transport projects as well as for the development of transport plans and SUMPs [24].

When designing inclusive mobility services and policies, it is important to first understand and then respond to a wide range of user needs. All the potential end-user groups should be engaged in the design, test and evaluation of the mobility solutions in order to maximise inclusivity.

There are some barriers to involving stakeholders and citizens successfully, often related to limited financial and personnel capacities within local authorities and also lack skills on how to plan and carry out a participation process and selecting the most appropriate involvement tools. Consultation processes can be long and time-consuming and "consultation fatigue" can be an issue [25]. Although it is a complex topic and several questions about participation still remain unanswered, citizen and stakeholder engagement are a prerequisite for long-term urban mobility planning [26]. The appointment of suitable governance structure at a horizontal level, for instance with the creation of a taskforce dedicated to this purpose is generally recommended[27, 28].

3.4 Political Support

Political support and leadership constitute important enabling factors for innovation and lay the foundation for change towards more sustainable urban mobility[9]. It is also the glue between the establishment of regulations and collaborative measures with citizens.

Long-term commitments and vision (sometimes accepting the risk to fail in some projects or trials) are important features driving innovative transformation forward. In this process a stable, supportive policy environment is a pre-condition for the uptake of new mobility solutions.

Coherent mobility plans over time can be facilitated by the adoption of the Sustainable Urban Mobility Planning process that embeds the creation of clear strategies and detailed implementation roadmaps. Long-term plans as is the case for the SUMPs enable the development of the urban mobility capacities required to have a long-term impact, beyond isolated initiatives.

Political support is also required to steer consumer behaviour towards more sustainable mobility options, maximising the adoption of these innovations [29]. Pricing and taxation are two widespread policies that can be used to promote sustainable mobility [30]. A comprehensive and systemic approach towards change implementation in urban mobility includes combining such demand-side initiatives with supply-side assistance for the development of new solutions.

As already pointed out in the previous sections, this needs to be complemented by more agile and flexible administrative procedures to facilitate the approval and deployment of urban mobility innovations, allowing regulatory sandboxes. A regulatory sandbox is *"a defined space where new business models, technologies and policies can be*

---

[9] https://www.eiturbanmobility.eu/wp-content/uploads/2020/12/122020_Urban-Mobility-Next.pdf.

deployed and used in a way that is safe and responsible[10]. Sandboxes support innovation in the cities and help policy makers to better understand the impact of new mobility solutions.

Likewise, clear and transparent communication and coordination with stakeholders and citizens are crucial in building the necessary consensus and delivering successful and scalable pilots leading to real-world transformations.

3.5   Access to the Right Funding Solutions

The urban mobility environment is highly dynamic. This feature makes it very attractive to major investments. However, the distribution of funding is highly uneven, concentrated on specific business models and on a few individual companies outside the EU. Innovators need access to the right financing solutions to test and scale up new products and services. Public budgets are limited and investments in infrastructure and transport services compete against other spending priorities, and private investors are often reluctant to invest in sustainable transport projects. Thus, cities need to seek additional funding and financing options and to develop business models to attract private sector investments in the development of the urban transport system. As a result, cities must explore additional funding and finance sources, as well as establish business models to attract private sector participation in the development of the mobility system [31]. Figure 4 shows an overview of funding and financing options for sustainable urban transport measures.

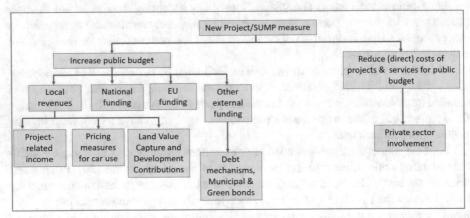

Fig. 4.   Overview of funding and financing instruments [23]

The EC provisions direct funding grants from its executive agencies for projects with specific objectives. Main programmes are[11]:

---

[10] https://www.gov.uk/government/consultations/future-of-transport-regulatory-review-regula tory-sandboxes/future-of-transport-regulatory-review-regulatory-sandboxes#definition.
[11] https://www.eumayors.eu/support/funding.html.

- Connecting Europe Facility (CEF), created to accelerate the development of transport infrastructure across the EU
- Horizon Europe (2021–2027 Research and Innovation programme). HE incorporates 5 research and innovation missions. Mission areas which are relevant for urban mobility include 'Climate-neutral and smart cities' and 'Adaptation to climate change including societal transformation'.
- Innovation fund, one of the world's largest funding programme for the demonstration of innovative low-carbon technologies
- The LIFE Climate change mitigation and adaptation Programme
- URBACT is a European exchange and learning programme promoting sustainable urban development
- Interreg: European Territorial Co-operation (cross-border, trans-national and interregional)
- The Renewable Energy Financing Mechanism (REFM) to support renewable energy projects, enable EU countries to work more closely together and achieve both their respective and collective renewable energy targets

To fulfil the objectives of the European Green Deal, it is critical to identify bottlenecks and barriers to innovation and market development. The field of urban green mobility solutions and services is today mostly dominated by non-European start-ups. This is mainly due to a better access to equity financing for non-European companies and the existence of more difficulties in scaling up in Europe due to heterogeneous markets with regards to policies, legislation and regulation.

Thus, in order to remove the obstacles identified and improve access to financing for innovative transport companies in European cities, the following recommendations should be followed [31].

- Incentivise Public Transport Operators and Authorities to open up to third party digital mobility platforms
- Introduce a clear and standardised EU-wide definition and regulation of mobility services
- Tailor flexible grants for fast growing service companies

As mentioned above, the development and implementation of new innovative mobility solutions require considerable investments that are difficult to fund with traditional public finance. In this context, Public-Private Partnerships (PPP) can be a very useful solution to overcome the shortage of public finance and cuts on public spending.

The establishment of PPPs is a method of long-term cooperation of public and private sectors in the implementation of projects aimed at the provision of public services. It allows the public sector to obtain resources from the private sector through a contractual agreement. This financing mechanism secures funding for the overall life cycle of the project. The aim of the cooperation is to achieve optimum performance of a public service and mutual social and commercial benefits between the parties [32].

Europe is increasingly deploying large-scale demonstrators and small-scale testing units that adopt a PPP approach. These long-term agreements typically include [33]:

- A financial commitment from both sectors public and private;
- The deployment of the demonstrator or the testing unit by the private sector for a given period of time;
- The commitment of the public sector to being a facilitator for demonstration and testing activities, whether in terms of political support or the provision of infrastructure by municipal authorities; and
- The sharing of the risk-reward potential derived from delivering the services or infrastructure.

Such cooperation between the public and private sectors enables businesses to industrialize and validate their innovations. Likewise, enables private sector to commercialize and profit innovative solutions. On the other hand, public sector is able to boost regional competitive advantages that lead to economic growth and create quality jobs.

4   Guiding Cities Through Transition

With the aim to guide cities through their urban mobility transition, and thus develop effective policy responses to emerging solutions, the hereby proposed approach focuses on understanding the impacts as well as the operational feasibility of such new mobility solutions.

In the context of the Sustainable Policy Response to Urban Mobility Transition (SPROUT) project that is funded by the Horizon 2020 programme, an evaluation framework [34] was developed which is structured around two main pillars: operational assessment of the pilot impacts (outcome evaluations) and assessment of urban mobility policy responses in the pilots (process evaluation). For both pillars, the evaluation tackled the following:

- Methods for performing the assessment
- Assessment indicators
- Information needed from use cases or other sources
- Information collection means and sources
- Limitations

In order to do so, it built upon a combination of existing methodologies, among others FESTA methodology for assessing Field Operational Tests (FOTs), Cost Benefit Analysis (CBA) for financial and economic aspects of the pilots, Global Logistics Emission Council (GLEC) methodology on emissions reporting for environmental impact, as well as specific CIVITAS tools, such as the NISTO evaluation toolkit or multi-actor multi-criteria analysis (MAMCA).

4.1   SPROUT Project Evaluation Framework

The present paper details the insights from the application of the framework to six cities in the framework of H2020 SPROUT project.

SPROUT implemented nine pilots in five cities focusing on feasible and sustainable emerging mobility solutions that could benefit from an appropriate policy response. This policy response could translate into improved sustainability or decreased negative impact of the solution.

- This was done following a three-pronged approach:
- test the new mobility solutions and assess the operator's operational feasibility and financial sustainability
- assess the economic, environmental & social impacts of the new mobility solutions and identify potential policy intervention areas
- assessing policy-related and regulatory barriers

As mentioned above, the pilots tested that an appropriate urban policy response could be implemented to harness the benefits of the emerging mobility solutions. In order to do so, local policymakers were involved jointly with other relevant stakeholders to prioritise the potential policy responses, and a subset of those were introduced in a limited scale. The proposed evaluation framework not only enabled assessing their implementation feasibility but also their user acceptance.

Stakeholder workshops and surveys were used for assessing the urban mobility responses in each pilot city. On top of that, SPROUT leveraged the commonalities among the pilots, to gain a deeper insight into their outcomes. This led to key policy implementation messages accompanying the successfully tested city policies.

Indeed, when looking at potential policies and user acceptance, it was important to understand that the adoption of new urban mobility solutions require defining policies that not only target city goals (e.g. reduce environmental impact) but also do not worsen other variables (e.g. accidents).

This depends on the city stakeholders? levels of acceptance. While service operators focus on ensuring operational feasibility and financial sustainability of the solution; the city targets maximal social and environmental benefits with an associated minimal cost. In any case, citizens are determinant for adoption success, as they represent both end-users who benefit from the service and/or those who bear the consequences.

The role of policymakers is key in catalysing all the stakeholders' requirements by defining tailored policies to each specific idiosyncrasy. However, as stated throughout this paper, mobility solutions emerge fast, leaving them little room for reaction. Thus, a policy evaluation framework as the one proposed in this chapter would improve their decision-making process, not only guiding them with a clear methodology but also relying on fact-based evidence.

Indeed, the application of the proposed evaluation framework in SPROUT project gravitated around the already mentioned three-pronged approach that can be seen in Fig. 5. The SPROUT project adapted the generic FESTA methodology[12] for planning and running a field operational test to cover the pilots' activities, from guiding their setup to appraise the outcome and the process. Thus, evaluation is divided into three phases:

---

[12] https://cordis.europa.eu/project/id/610453.

- Preparing phase: It focuses on the definition of the research questions that will help to find the indicators and define the collection and assessment methods pilots will use during the next phases.
- Using phase: It covers the data collection phase when using the mobility solution and performing user acceptant test, questionnaires and workshops.
- Analysis phase: Analyse the compiled data to define the policies requiring intervention or being removed and draw the city policy response.

"Cross-cutting issues" are depicted in the centre of Fig. 5. They include all the aspects considered by the FESTA methodology such as the implementation plan & context definition; the role and involvement of the stakeholders that will participate in the pilot activities; the ethical and legal issues required for ensuring data privacy, and cultural or regional backgrounds. As pilots are small-scale multi-stakeholder demonstrators, it is essential they define the communication strategy and foresee any event that may disrupt the initial implementation plan. Therefore, the SPROUT project included two additional aspects: communication strategy and risk management.

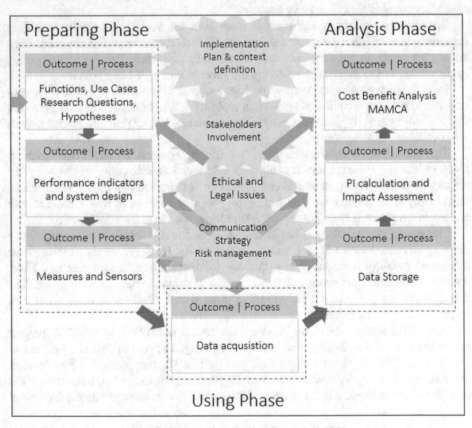

Fig. 5. Proposed evaluation framework [34]

Regarding implementation, on a first stage, pilots tested in practice the emerging mobility solutions, introducing them into a limited scale "real ecosystem". There, they collected data that not only enabled assessing the operators' operational feasibility and financial sustainability, but also the sustainability impact. Building upon these data, cities identified policies that had the potential to enhance these results by being modified or removed. The evaluation framework responded to questions on how to measure operators' sustainability and operational feasibility, and the sustainability impact of the new mobility solution, as well as how to use the data-based evidence to identify the policies which should be modified or removed.

The second phase focused on the resulting policies with negative impacts, the existing alternative responses and the compiled stakeholders' preferences, so pilots could evaluate and prioritize policies to incorporate. Again, the evaluation framework guided the cities on how to evaluate and prioritize the policies. In this regard, the selected methodology was multi-criteria analysis (MAMCA), which allows prioritizing the policy responses shows the synergies and conflicts between the stakeholder groups and determines the level of consensus of each alternative [35].

Last, from the list of prioritized responses, pilots' policy-makers selected a subset to be implemented at a limited scale. Pilots thus assessed their implementation feasibility and user acceptance to validate the alternative policies. This led to cities drawing the city-specific policy response. Finally, the evaluation framework provided a methodology on how to define and assess the implementation feasibility and user acceptance, as well as how to use the results for defining the final city-specific policy response.

5   Conclusions and Path Ahead

Impacts of emerging mobility solutions are inadequately addressed by current urban policies, as a successful transition requires collecting, managing, and sharing data. This is even more challenging in the case of urban freight policy making, due to lack or insufficient data accessibility.

This transition relies on the combination of several factors, with new consumption patterns stemming from economic growth and societal changes on the demand side, and digital technology advances together with widespread Internet access on the supply side. Transport policy plays a major role in this, not only by means of providing funding to transport investments but also by deploying physical, regulatory and pricing measures, along with promoting knowledge sharing and education.

Key enablers for urban mobility transition are an existing innovation ecosystem, quality data availability and analytical skills, citizens and stakeholders' engagement, political support, and access to funding, on top of local authorities' capacity to prepare and implement urban mobility strategies. Involving all potential user groups in the design, test and evaluation of mobility solutions is crucial in order to ensure inclusivity and accessibility.

This chapter proposes an approach for guiding cities towards the implementation and adoption of new digital urban mobility solutions. Specifically, this paper proposes an evaluation framework to guide cities assess the operational outcomes of pilots as well as the urban mobility policy responses (processes) in those pilots. These two pillars (outcomes and processes) are intertwined, as a successful evaluation is essential for both, i.e. running implementations and testing activities smoothly; and assess the impacts that ultimately support decision-making. This evaluation framework is meant to be used by any city that wants to speed up the policies definition when introducing new mobility solutions.

As next steps, this chapter outlines how the above proposed evaluation framework could be complemented with the insights stemming from H2020 INDIMO project on inclusive digital mobility solutions. Indeed, INDIMO evaluation framework [36] incorporates inclusivity and accessibility among its building blocks. This could be added as a third pillar to the outcomes and processes ones hereby presented, not only ensuring that urban mobility transition is inclusive and accessible, but also that it minimizes physical, cognitive and cultural barriers, incorporates the gender perspective and tackles vulnerable groups' needs. Given the complementarity, combining the two projects outcomes and learnings would help cities to balance the triple-bottom line of the sustainability.

In summary, by ensuring a correct execution of the proposed evaluation frameworks, cities can draw city-specific policy responses to ensure the satisfactory adoption of new mobility solutions. Therefore, the present evaluation framework lays the grounds for guiding policymakers through inclusive urban mobility transition. The scalability potential of the proposed frameworks can play a key role in the overall transition to climate-neutral economies and societies.

This is indeed one of the challenges recognized by the European Commission in its Research and Innovation Programme Horizon Europe[13]. In order to address them, the programme focuses on supporting and implementing EU policies with open calls addressing desired impacts – that the EC refers to as destinations.

Already the first calls of Horizon Europe programme present destinations that are building upon the grounds laid in this paper. Indeed, the *Connected, Cooperative and Automated Mobility (CCAM)*[14] destination can leverage the hereby presented outcomes evaluation strategy by calling for "*all technologies, solutions, testing and demonstration activities being documented fully and transparently, to ensure replicability, increase adoption, up-scaling, assist future planning decisions and EU and national policy-making and increase citizen buy-in*"[15].

Moreover, the Cross-sectoral solutions for the climate transition destination expects the engagements of citizens and stakeholders, in line with the process evaluation presented in this paper. Indeed, this destination targets, among others, "more effective policy interventions, co-created with target constituencies and building on high-quality policy advice" and "greater societal support for transition policies and programs, based on greater and more consequential involvement of those most affected" (See footnote 15). Projects addressing this destination could therefore build upon the hereby presented evaluation approach, with its specific stage to compile stakeholders' preferences, that are subject to later evaluation and prioritization by policymakers.

---

[13] https://ec.europa.eu/info/research-and-innovation/funding/funding-opportunities/funding-programmes-and-open-calls/horizon-europe_en.

[14] https://www.ccam.eu/what-is-ccam/ccam-partnership/.

[15] https://ec.europa.eu/info/funding-tenders/opportunities/docs/2021-2027/horizon/wp-call/2021-2022/wp-8-climate-energy-and-mobility_horizon-2021-2022_en.pdf.

Beyond this specific destination, future research will involve a broader integration of citizen and stakeholder engagement across the whole Horizon Europe programme.

Acknowledgments. This chapter has been developed in the framework of the European Project "SPROUT: Sustainable Policy Response to Urban Mobility Transition", funded by the European Commission under the European Union's Horizon 2020 research and innovation program. Grant agreement No 814910.

This paper reflects the views only of the authors and not the official opinion of the European Commission.

References

1. Future of mobility: the UK freight transport system. British Government Office for Science (2019). https://assets.publishing.service.gov.uk/government/uploads/system/uploads/attachment_data/file/777699/fom_freight_sharing_economy.pdf
2. Hauben, H., Lenaerts, K., Waeyaert, W.: The platform economy and precarious work, Publication for the committee on Employment and Social Affairs. Policy Department for Economic, Scientific and Quality of Life Policies, European Parliament, Luxembourg (2020). https://www.europarl.europa.eu/RegData/etudes/STUD/2020/652734/IPOL_STU(2020)652734_EN.pdf
3. Jones P.: The evolution of urban mobility: the interplay of academic and policy perspectives. IATSS Res. **38**(1), 7–13 (2014). ISSN: 0386–1112. https://doi.org/10.1016/j.iatssr.2014.06.001
4. UITP. The Future of Mobility post-COVID. Union Internationale des Transports Publics. & Arthur D. Little Future Lab (2020). https://www.uitp.org/publications/the-future-of-mobility-post-covid/
5. Te Boveldt, G., Tori, S., Keseru, I.: SPROUT D2.3 Urban Mobility Transition Drivers (2021)
6. European Court of Auditors (Special Report, 2020). Sustainable Urban Mobility in the EU: No substantial improvement is possible without Member States' commitment (2020). https://op.europa.eu/webpub/eca/special-reports/urban-mobility-6-2020/en/
7. Kukely, G., Attila, A., Fleischer, T.: New framework for monitoring urban mobility in European cities. Transp. Res. Procedia. **24**, 155–162 (2017). https://doi.org/10.1016/j.trpro.2017.05.081
8. Head, B.: Evidence-based policy: principles and requirements. Strengthening Evid.-Based Policy Aust. Fed. **1**(1), 13–26 (2010)
9. Governing better through evidence-informed policy making. Conference summary, OECD (2017)
10. Addanki, S.C., Venkataraman, H.: Greening the economy: a review of urban sustainability measures for developing new cities. Sustain. Cities Soc. **32**, 1–8 (2017). ISSN: 2210–6707. https://doi.org/10.1016/j.scs.2017.03.009
11. May, A., Boehler-Baedeker, S., Delgado, L., Durlin, T., Enache, M., van der Pas, J.-W.: Appropriate national policy frameworks for sustainable urban mobility plans. Eur. Transp. Res. Rev. **9**(1), 1–16 (2017). https://doi.org/10.1007/s12544-017-0224-1
12. Royo, B., de la Cruz, T., Sánchez, S.: SPROUT D4.14: Policy implementation messages from cross-pilot results (2022)

13. Tsavachidis, M., Le Petit, Y.: Re-shaping urban mobility – Key to Europe´s green transition. J. Urban Mobility **2**, 100014 (2022). ISSN: 2667–0917. https://doi.org/10.1016/j.urbmob.2022. 100014

14. Mazzarino, M., Masetto, C., Cicarelli, G., Rubini, L., Coin, L.: SPROUT D4.5 Impact assessment and city-specific policy response: Padua Pilot (2022)

15. Xenou E., Touloumidis D., Aifantopoulou G.: SPROUT 5.2 Urban policy system dynamics model (2022)

16. Tori, S.M., Te Boveldt, G., Keserü, I., Macharis, C.: SPROUT D3.4 SPROUT Narrative scenario (2021)

17. EC (2017) Indicators and Data Collection Methods on Urban Freight Distribution, Non-binding guidance documents on urban logistics, no 6/6

18. Van Duin, J.H.R., Quak, H.J.: City logistics: a chaos between research and policy making? A review. In: Brebbia, C. (Ed.), Urban Transport and the Environment in the 21st Century, pp. 135–146 (2007)

19. Rubini, L., Della Lucia, L.: Governance and the stakeholders' engagement in city logistics: the SULPiTER methodology and the Bologna application. Transp. Res. Procedia **30**, 255–264 (2018). ISSN: 2352–1465. https://doi.org/10.1016/j.trpro.2018.09.028

20. La estrategia Española de movilidad sostenible y los gobiernos locales, Federación española de municipios y provincias, Red española de ciudades por el clima, Gobierno de España (2010) https://www.redciudadesclima.es/sites/default/files/2020-06/2a7fb70e4f9cfdd19fbd0 5d0240327b0.pdf

21. Fioretti, C., Pertoldi, M., Busti, M. and Van Heerden, S.: Handbook of Sustainable Urban Development Strategies, EUR 29990 EN, Publications Office of the European Union, Luxembourg, JRC118841 (2020). ISBN 978-92-76-13673-6. https://doi.org/10.2760/32842

22. European Commission, 2013. A Concept for Sustainable Urban Mobility Plans, Annex to Communication from the Commission to the European Parliament, the Council, the European Economic and Social Committee and the Committee of the Regions, Together towards competitive and resource-efficient urban mobility, Brussels, 17.12.2013, COM(2013) 913 final, 12.06.18

23. Topic Guide Funding and Financing of Sustainable Urban Mobility Measures (2019). https:// www.eltis.org/sites/default/files/funding_and_finance_of_sustainable_urban_mobility_m easures.pdf

24. Rupprecht Consult. Why is Participation a challenge in sustainable urban mobility planning? (2013)

25. Val, S., de la Cruz, T.: CIVITAS-SUNRISE D3.2 Cluster Recommendations: Innovative Solutions to Urban Logistics (2020)

26. Lindenau, M., Böhler-Baedeker, S.: Citizen and stakeholder involvement: a precondition for sustainable urban mobility. Transp. Res. Procedia **4**, 347–360 (2014). ISSN: 2352–1465 https://doi.org/10.1016/j.trpro.2014.11.026

27. Marciani, M., Cossu, P., Pompetti, P.: How to increase stakeholders' involvement while developing new governance model for urban logistic: turin best practice. Transp. Res. Procedia **16**, 343–354 (2016). ISSN: 2352–1465. https://doi.org/10.1016/j.trpro.2016.11.033

28. Guimarães Pereira, A., Cuccillato, E., Figueiredo Nascimento, S., et al.: Citizen engagement in science and policy-making, Publications Office, European Commission, Joint Research Centre (2016). https://data.europa.eu/doi/10.2788/40563

29. Ricci, L., et al.: Incentivizing sustainable mobility through an impact innovation methodology. CERN IdeaSquare J. Exp. Innov. **4**(2), 25–29 (2020). https://doi.org/10.23726/cij.2020.1055

30. Gallo, M., Marinelli, M.: Sustainable mobility: a review of possible actions and policies. Sustainability **12**, 7499 (2020). https://doi.org/10.3390/su12187499

31. European Investment Bank: Financing innovation in clean and sustainable mobility Study on access to finance for the innovative road transport sector (2018)

32. Koscielniak, H., Górka, A.: Green cities PPP as a method of financing sustainable urban development. Transp. Res. Procedia **16**, 227–235 (2016). ISSN: 2352–1465. https://doi.org/10.1016/j.trpro.2016.11.022
33. Probst, L., et al.: PwC Luxembourg. Public Private Partnerships Large-Scale Demonstrators & Small-Scale Testing Units. European Union (2013)
34. Royo, B., de la Cruz, T.: SPROUT D4.1: Pilots Evaluation Framework (2020)
35. Macharis, C., Turcksin, L., Lebeau, K.: Multi actor multi criteria analysis (MAMCA) as a tool to support sustainable decisions: State of use. Decis. Support Syst. **54**(1), 610–620 (2012). https://doi.org/10.1016/j.dss.2012.08.008
36. Basu S., et al.: INDIMO D4.1 Evaluation Framework (2020)

Automated Vehicles Empowering Mobility of Vulnerable Groups - and the Pathway to Achieve This

Annette Randhahn[1]([✉]), Joana Leitão[1], Erzsébet Foldesi[2], Jörg Dubbert[1], Alexandra Pinto[1], Carolin Zachäus[1], and Peter Moertl[3]

[1] VDI/VDE Innvation + Technik GmbH, Berlin, Germany
{annette.randhahn,joana.leitao,joerg.dubbert,alexandra.pinto,
carolin.zachaeus}@vdivde-it.de
[2] Self-employed consultant on inclusion of disabled persons, Budapest, Hungary
[3] Human Centred Systems, Virtual Vehicle Research GmbH, Zurich, Switzerland
Peter.Moertl@v2c2.at

Abstract. Many people in Europe still have limited access to transportation modes overall. Socio-economic constrains as well as cognitive, sensory and physical impairments affect everyday life of these citizens, posing challenges to access mobility services.

Technologies for vehicle automation have advanced greatly in recent decades and it is expected to become part of vehicle fleets in the foreseeable future. Yet, the implementation and use of automated and autonomous vehicles (here jointly referred to as AVs) entails chances but also hurdles regarding accessibility and inclusivity of vulnerable groups. This concerns both the use of the vehicle by humans as well as the interaction between humans and vehicles as participants in road traffic.

In this chapter, these aspects shall be presented by identifying opportunities and risks as part of our mobility system, starting from a narrowing down of the vulnerable social groups we are looking at. Subsequently, we present the benefits that co-creation and universal design can have in overcoming or, in the best case, avoiding these obstacles. Even though the authors are aware that no detailed recommendations for action can be given within this framework, at least suggestions for solutions are outlined.

1 Introduction

Driverless vehicles are no longer something just anticipated by visionaries, confined to research labs and a technology of the future. In fact, many new vehicles on the road are already being equipped with automation level 2 functions (overview of automation levels see Table 1). Most recently, Mercedes (2022) received a lot of attention, as it is now the first car manufacturer in the world to offer an approved Level 3 system and is thus also liable for accidents during automated driving mode. Furthermore, we are all familiar with the automated and autonomous vehicles (here jointly referred to as AVs) used for Google Street View (Reuters 2020) or the autopilot system developed by Tesla (2022) that has already made headlines in every respect. Both these examples are already driving on the road, wherever the legal framework permits. In contrast to extensive discussions on the environmental effects, legal issues and safety aspects of this technology, the questions of accessibility and inclusivity have so far only been dealt with in an insufficient manner.

A study conducted by Neumann et al. (2003) found that 48.1% of people with disabilities in Germany would travel more frequently if there were more barrier-free options. While some barriers have been removed since then, people with disabilities continue to travel significantly less than those without disabilities and experience notably more travel difficulties with any type of trip (Clery et al. 2017).

And yet, the access to society and transport for people with disabilities receives a high level legal obligation of countries. The United Nations Convention on the Rights of Persons with Disability (UN CRPD) was adopted in 2006 and came into force in 2008. The UN CRPD does not explicitly mention the right to access AVs for people with disabilities but includes articles that cover the obligation to provide access for disabled users, on equal basis with others, to transportation and technologies[1]. The EU and its Member States have ratified the Convention, and therefore undertaken such obligations. On EU level, the Articles 25 and 26 of the Charter of Fundamental Rights of the European Union (GRC 2012) include related issues, i.e. the rights of the elderly and integration of persons with disabilities and their rights to lead an independent life whereas access to transport is a precondition.[2] The existence of both of these articles emphasises the importance of including vulnerable groups in society and facilitating their autonomy and independence.

How this can be supported with the help of AVs will be outlined in this chapter. In doing so, we draw not only on current scientific literature, but also on the findings of the research project HADRIAN, which will be briefly presented below.

The project Holistic Approach for Driver Role Integration and Automation Allocation for European Mobility Needs (HADRIAN)[3] aims to shape automated driving from a holistic, user-centred perspective, starting the development process with specific mobility scenarios concerning individual users as well as mobility needs and constraints.

---

[1] Article 3 - General principles: defines several general principles, including 'full and effective participation and inclusion in society' and 'accessibility'. These general principles are to be considered as preconditions to exercise all rights included in the Convention. Article 9 – Accessibility: requires State Parties of the UN CRPD to take appropriate measures 'to ensure to persons with disabilities access, on equal basis with others, to the physical environment, to transportation, to information and communication, including information and communication technologies and systems, and to other facilities and services open or provided to the public'. Article 20 - Personal mobility: includes an obligation on the States Parties to take effective measures to ensure personal mobility with the greatest possible independence for persons with disabilities'.

[2] Article 25 - The rights of the elderly: The Union recognises and respects the rights of the elderly to lead a life of dignity and independence and to participate in social and cultural life. Article 26 - Integration of persons with disabilities: The Union recognises and respects the right of persons with disabilities to benefit from measures designed to ensure their independence, social and occupational integration and participation in the life of the community.

[3] https://hadrianproject.eu/.

This much more technical project offers us insights regarding the interfaces for human-technology interaction specifically for older people, which are at least partially transferable to other vulnerable groups as well. In addition, a co-determination approach was also part of the methodology in this project.

But before focusing on the potentials and the risks of AVs for citizens from vulnerable groups in mobility, the basics of automated driving are addressed in Sect. 2. In addition to an explanation of the automation levels, we approach the following questions: Where do we stand technologically and will we all soon be sitting passively in self-driving cars and buses? Sect. 3, after defining the vulnerable groups addressed in this chapter, presents the potential benefits, as well as new barriers, that the use of AVs in our mobility system can bring to these vulnerable groups. Section 4 will show how the previously mentioned hurdles can be overcome. In this respect, the potential of the use of universal design will be considered in particular. A good practice example of automation in vehicle development provides Sect. 5. The HADRIAN project will be used to illustrate how automation levels 2 and 3 can facilitate the activities of the vulnerable elderly group in practice. Section 6 contains the conclusions of the previously identified aspects.

2 Automated and Autonomous Driving – A Glance at the State of the Art

2.1 Level Classification of Automated Driving

The terms *automated* and *autonomous* are often interchanged in non-scientific representations and discussions, although they do not mean the same thing. In technical and scientific literature, different levels of automated driving are mentioned. Depending on the defining instance, the number of levels and subtleties in the respective descriptions vary. In order to give an introduction to the basics of autonomous driving, we will use the latest standards of SAE International (2021). According to them, the levels of automation can be defined as follows:

According to this classification, levels 0 to 2 of automation are actually the major fraction of vehicles already on the road today. Vehicles with technology assisting the driving experience and contributing to increase road security such as ABS and ESP functions required by law, or assisted parking and acceleration tools, are indeed classified as level 1 and level 2 automated vehicles, respectively. Although not explicitly mentioned in this case, following the classification presented in Table 1, autonomous vehicles are those belonging to level 5, i.e. those that are in fact driverless vehicles.

2.2 Current Discussions on AVs Barriers - Why Do not Cars Already Drive Autonomously?

While AVs can transport both people and goods, the notable AVs currently in fast expansion and with potential for deployment in near future are robot-taxis, bus shuttles or similar forms of public transportation. Currently there are already level 3 public transport shuttles driving on small and controlled sections of roads in different cities, never without the accompaniment of a human driving assistant and mostly on pre-programed

Table 1. Level of automation in vehicles (SAE International, 2021)

Level	Term	Description
Level 0	No driving automation	Purely manual vehicle guidance, driver support features are limited to warnings and momentary assistence, e.g., blind spot warning or emergency brakes
Level 1	Assisted driving	The driver has full control of vehicle but is assisted by features that provide steering *or* brake/accelaration support
Level 2	Partial automation	The driver has to maintain monitoring and full responsibility of the vehicle but is assisted by features that provide steering *and* brake/accelaration support
Level 3	Conditional automation	The driver has to reengage driving when requested, but the AV has lateral and longitudinal control under limited conditions, e.g., traffic jam chauffeur or automated parking
Level 4	High automation	The vehicle drives itself under limited conditions that are much broader than on level 3, e.g., local driverless taxi. The driver is not necessary anymore, but may transition the vehicle to manual driving under some conditions
Level 5	Full automation	The vehicle can drive alone on all level of complexity for driving conditions. All occupants are passengers

routes. Martinez-Diaz and Soriguera (2018) provide an overview of AV technology and challenges that are still an issue.

Although the technology of automated driving functions has grown rapidly in recent years, current forecasts of market availability are no longer as optimistic as those from a few years ago. Some manufacturers have already presented prototypes for level 4 and 5 cars, trucks (Volvo 2022), bus shuttles or even buses (CAVForth 2022). However, most expert predictions do not foresee a generalized availability of high or fully automated vehicles on the market and on-road fleet in the near term. Roos and Siegmann 2020, state in a technology roadmap based on numerous expert interviews that since the technological and also regulatory leap from SAE level 3 to high level of vehicle automation (SAE level 4) is significantly higher than from level 2 to 3, highly automated driving will only be possible between 2040 to 2050. The current obstacles are numerous in quantity and kind. The challenges can be divided into technical and non-technical ones, inspired by the 5-layer model on automated driving (Eckstein 2016). Besides technical aspects (layer 1) the non-technical challenges (layers 2–5) human factors, economic, legal and societal aspects play a major role. These different layers are interlinked with each other and cannot be regarded strictly separately.

Starting with the technical issues, recent Research and Development (R&D) and demonstration projects like HEAT (2020) or STIMULATE (2021) have shown that there are still several technical limitations to be improved. Being able to deal with deviations from the programmed route, e.g. due to road works, dealing with unpredictable road obstacles or being able to drive in all-weather conditions are just a few of them. Liu et al. (2020) state, that there is still a remarkable gap between current state of the art of computing and communication systems and the expected robust system for level 4 and 5 autonomous driving.

Furthermore, implementation of AVs will bring new cyber risks with it, such as hardware and software failures and potential hacking attacks (Litman 2022), which are directly related to societal factors and people's concerns regarding their safety, privacy and data protection (Fagnant and Kockelman 2015).

From an economic point of view the co-existence of automated and non-automated vehicles on the road is a challenging scenario. Not only do the vehicles have to be equipped with expensive sensor technology, algorithms and chip technologies (which, in addition have recently become difficult to purchase) but the infrastructure also requires expensive recognition and communication technology. Especially in the complex urban environment, the cost factor plays a significant limiting role for the implementation of autonomous vehicles and public transport services.

Very challenging legal aspects still to be solved are security and liability in case of accidents. AVs are ultimately just machines that will be programmed by humans and all decisions made on the road will be based on pre-defined guidelines. Two equally programmed machines would unlikely be involved in the same accident, but when the roads are shared with other users such as cyclists and pedestrians, the scenario becomes more complex, with many unpredictable variables. Currently, there is no societal consensus on how such ethical and moral as well as liability guidelines should be enforced on the vehicles and how accidents should be addressed. It is in any case vital to ensure that the roads can be safely shared among different users. In spite of Germany having already published its ethical guidelines (BMVI 2017), these are not acknowledged across borders. To further complicate matters, machines are susceptible to functioning errors or could even potentially fall victims of cyber-attacks, which would undermine the road safety even if it were well defined from the start. However, there is much progress to develop or update and implement regulations addressing security as well as liability of AVs (UNECE 2020).

One considerable societal impact that needs to be addressed is in fact common to many sectors profiting from digitalization. Replacement of humans by different machines and/or robots will deem some professions obsolete leading to an increase of unemployment. In the case of autonomous vehicles, the redundancy of drivers and other functions in the transport sector might hinder social acceptability of these vehicles. On the other hand, there is a considerable shortage of skilled workers, not only in the transport sector, so that the elimination of the need for a driver could be minimized when addressed with suitable capacity building opportunities.

As mentioned before, the different layers of automated driving are interlinked and need to be addressed in a holistic approach. AVs are attributed with some advantages, such as improved and more energy-efficient driving and consequently the reduction of congestion and road mortalities (see e.g. Krail et al. 2019). However, it is important to emphasize that not all scenarios are as positive regarding these benefits and that AVs can potentially lead to an increase in car-use due to low occupancy rates and travelled distance, and therefore, an increase in energy use (e.g. Acheampong et al. 2021). Therefore, the implementation of a policy framework of governance measures is necessary to ensure that the environmental potentials are exploited and the corresponding risks are minimized. To give an example, without efficient road pricing it may be cheaper for users to have their vehicles to continue driving around the block instead of parking which would add to traffic congestion.

Last but not least, an important factor for the adoption of AVs will be the acceptability by the population in general, which is affected by aspects of all layers such as attractiveness, cost and trust in new technology and appropriate regulatory frameworks, and also the named burden on the environment. Penmetsa et al. (2019) presents a summary of studies focused on public perceptions of AVs. Because of the identified challenges and often for purely psychological reasons, many people simply cannot yet imagine being able to sit back, relax, work or even sleep while being driven and completely relinquishing the driver's role to the car. This resistance to change may become the main limiting factor once technology has advanced sufficiently. So, while there is a major technology development taking place, society as a whole may not be ready to have vehicles above automation level 4 on the road just yet. The same applies to other road users, who potentially do not feel safe in the presence of driverless vehicles, although the human driver has significantly more sources of error, such as mistaking the accelerator and brake, falling asleep at the wheel or driving drunk. In addition, interaction with other road users has already become much safer thanks to driver assistance systems. One example is the turn-off assistant for trucks, which warns the driver of pedestrians and cyclist in blind spots (Weinrich 2017). Hence, for a successful deployment of full AVs, it is important that the general population is introduced to the use of vehicles where no human will be in control.

Against the backdrop of these multifaceted discussions in science, politics and among the public, and although the societal factor is giveen more and more priority, aspects regarding inclusivity and access to mobility for vulnerable groups are hardly considered. Before we go into this in more detail, however, we should define which vulnerable groups we want to take a closer look at here.

3 How AVs Can Enhance the Mobility of Vulnerable Groups

3.1 Narrowing Down the Definition of Vulnerable Groups

There is currently no clear and established definition of vulnerable groups used uniformly by international organizations and authorities. While vulnerable groups are often considered as those at risk of poverty and social exclusion or with some type of disability, a more comprehensive definition was proposed in the European Recast Reception Conditions Directive (2013). Art. 21 of the directive defines vulnerable persons as:

(...) minors, unaccompanied minors, disabled people, elderly people, pregnant women, single parents with minor children, victims of trafficking in human beings, persons with serious illnesses, persons with mental disorders and persons who have been subjected to torture, rape or other serious forms of psychological, physical or sexual violence, such as victims of female genital mutilation (...).

While the Inter-agency Network for Education in Emergencies also includes this definition, they provide a more comprehensive definition by stating: "Vulnerable groups are physically, mentally, or socially disadvantaged persons who may be unable to meet their basic needs and may therefore require specific assistance. Persons exposed to and/or displaced by conflict or natural hazard may also be considered vulnerable." (UNHCR 2006).

In addition, there are several pieces of EU legislation on passengers' rights in the field of transport including the rights of passengers with reduced mobility and disability. These legislations, e.g. the EU regulation on rail passengers' rights and obligations (European Parliament 2021) gave a human rights based definition on persons with disabilities and those with reduced mobility saying that:

A 'person with disabilities' and a 'person with reduced mobility' mean any person who has a permanent or temporary physical, mental, intellectual or sensory impairment which, in interaction with various barriers, may hinder his or her full and effective use of transport on an equal basis with other passengers or whose mobility when using transport is reduced due to age;

This definition is significant as it draws the attention to the fact that disability results from the interaction between persons with impairments and various physical, information-communication etc. barriers. In terms of autonomous cars, it means that if persons with disabilities are considered when designing AVs, they could be able to use these cars on an equal basis with other passengers.

These definitions are in line with the vulnerable groups identified by the INDIMO project (Kedmi-Shahar et al. 2020) which include the groups in the table below with their corresponding specific needs.

Table 2. Vulnerable user groups, their share in European society and examples of their specific needs (Eurostat 1 2022, Eurostat 2 2022, Eurostat 3 2022, Eurostat 4 2022, Eurostat 5 2022, EU-KOM 2020, EU-KOM 2021, ENAR 2019, EBU 2022, Kedmi-Shahar et al. 2020)

Group	Share of European Society	Example for specific needs
Elderly people (65 or over)	19.2%	Simplified user interface
People with reduced mobility	5%	Announcement of obstacles on path
People with reduced vision	3.3%	Assistance in interacting with the environment
Women	51%	Strengthening of autonomy
People living in rural areas	29.1%	New mobility innovations
Foreign people	12.4%	Various language and payment options
Ethnic minorities	10%	Various language options
People with low income	21.9%	Affordable fares and multiple payment options, including cash payments
Caregivers of children	29.9%	Possibility to transport children in strollers

3.2   Meeting Gaps in the Mobility System - Benefits AVs Can Bring to Users and Vulnerable-To-Exclude Groups

Vaa (2003) shows that due to certain physical and cognitive impairments (e.g. visual, neurological, hearing, medical condition, mental workload, and easy distraction) driving capabilities might reduce with age. Furthermore, elderly people generally seem to have little self-assessment capabilities of their own driving abilities. A paper by Horswill et al. (2011) shows that elderly drivers have poor insight into their own hazard perception capabilities. With one fifth of the European population being over 65 (see Table 2), certain impairments can lead to severe safety limitations for the driver and other road users.

The HADRIAN project shows, even vehicles with SAE level 2 and 3 assistance systems can already take over essential driving tasks. That can enable certain vulnerable groups, like elderly people, to fulfil their own mobility needs, stay independent and at the same time are safe participants in road traffic. At the same time, persons who are too young to hold a driver's license could also benefit from an introduction of such systems: as the operation of a vehicle becomes easier, the legal age to use it may also be lowered. Furthermore, the detection of people suddenly stepping onto the road and the associated automatic braking contribute to improved overall safety, in particular of pedestrians with cognitive impairments and sensory limitations.

One to two SAE-levels higher, self-driving cars can enable independent mobility for people who temporarily or permanently cannot drive a car at all (elderly, people with intellectual or visual disabilities, epilepsy) or can only drive vehicles with major and expensive adaptations (severely physically disabled). Their dependency on these vehicles is exacerbated if their residency location, e.g. a rural area, does not provide public transport, or if that available is not accessible. Therefore, autonomous cars or buses may be important means of overcoming mobility barriers for those who currently face difficulties in driving or using cars.

During a workshop hosted by the Budapest Association of Persons with Physical Disabilities (MBE) representatives of organisations of persons with different disabilities have explicitly validated the above-mentioned potential of AVs and how they could transform their life by providing them with independent travelling and thus enabling them to actively take part in society (Földes 2017). A survey conducted by Földes et al. (2019) showed furthermore, that the importance of the presence of staff for future users of AVs is not an essential issue for disabled people. The respondents, from which 6% declared themselves to be mobility or visually impaired, did not identify the presence of staff as important, attributing only 1.6 on a scale from not important (1) to very important (3). Additionally, this was also rated as only slightly more important by mobility and visually impaired respondents (1.8).

In summary, that means, by eliminating the necessity of a human driver, people who cannot drive (fully) on their own will be able to move around more independently. This can take place in the private transport sector, for example, through the use of automated driving functions in private vehicles or also in the public sector through the increased offer of autonomous buses or taxis in rural areas, which currently only run a few times a day due to the personnel costs for drivers. In addition, the above mentioned shortage of skilled bus drivers also plays an increasingly important role for public transport operators.

Additionally, autonomous taxis and buses are expected to be a lower-cost transportation option, making them more affordable for non-drivers (Litman 2022). Thus, thanks to the introduction of AVs a multitude of users will become more mobile and able to reach more places (Rojas-Rueda et al. 2017). Meanwhile, they will be less dependent on assistance from other people in their everyday lives and can engage in society more freely. In the future, people could either own private vehicles that are able to drive autonomously or there could be shared vehicles that provide an on-demand service.

Consequently, people living in peri-urban or rural areas will be able to give up their personal cars. Furthermore, car-sharing services can become more flexible as users will no longer be required to pick up a car in a designated area. Instead, users will be picked up directly from their current location.

3.3 Potential Barriers and Risks of AVs for Vulnerable Groups

For all of their advantages, AVs also come with some disadvantages, particularly for certain vulnerable user groups. These groups are especially underprivileged in situations where people could be excluded due to financial means or other access limitations. Furthermore, there could be potential issues in the operation of automated assistance functions, in particular barriers related to the Human Machine Interface (HMI). In this context, people with little or no digital skills would not necessarily be able to use such services. Moreover, although it will likely still take years if not decades until vehicles are driving completely independently on our roads, there may be risks due to the remaining human drivers who tend to act more unpredictably.

In addition, AVs do not only have to interact with other drivers, but they must also avoid hitting moving and non-moving objects, including pedestrians which could be elderly people or people with low vision or reduced mobility who may not act the same or as fast as an able-bodied person would. Since AVs often already drive electrically, they can also be considered relevant for visually impaired people in terms of the associated quiet driving. Elderly people or people with visual and hearing disabilities and even pedestrians who use headphones are at risk of being hit by electric and hybrid cars due to how quiet they are at low speeds. Therefore, these cars must be fitted with an Acoustic Vehicle Alerting System (AVAS), a low-speed alerting sound to keep pedestrians safe. In the event of an accident, details regarding liability are still undefined (Fagnant and Kockelman 2015).

An additional aspect is the underlying infrastructure. In this regard, dramatic change is needed, especially regarding refuelling/charging and maintenance stations which need to be accessible for everyone and parking spots that must offer sufficient space for any user, including wheelchair users who may require ramps. So far, these aspects were not taken into account in city planning, resulting in an access barrier for certain vulnerable groups.

Moreover, there above mentioned the lack of trust in the new technology is particularly high among elderly people who are not used to technology taking over a task they used to undertake themselves. Diepold et al. (2017) found out that about 75% of elderly drivers are not willing to ride with automated vehicles due to uncertainty and distrust in the technology.

In order to address these challenges and avoid these barriers, it is important to identify the requirements of all users at an early stage. Then, the potentials mentioned above can be exploited.

4 How to Overcome the Barriers

4.1 Identifying Potential Requirements of Vulnerable Users When Using Autonomous Cars

The user of an AV must give information to the vehicle by the HMI on the one hand and receive information from it on the other. According to Földes (2019), for the information input, interfaces should be simple and accessible and not rely on one single sensory channel. It is important that control alternatives to vision, auditory, speech and tactile elements is provided, as, for example, visually impaired users would need both tactile interfaces and voice-controlled systems, but other disabilities might not allow for the use of systems that are exclusively voice-controlled. Furthermore, passengers with intellectual disabilities and autism are more reliant on support to navigate them from one place to another. However, for some people a system that offers the possibility for supervision and tracking the journey through video cameras and GPS can help their caregivers.

When AVs potentially share information with the passenger, it will happen either in terms of vehicle maintenance notifications or regarding the vehicle route (such as, current location, progress of ride, potential deviations, etc.). How differently this flow of information can be perceived is shown by a study by Kim et al. (2012) that investigated the impact of multimodal, in-vehicle navigation systems for different aged drivers. They found that, while young drivers benefit from multi-modal navigation systems, older people are oftentimes overwhelmed because of their already high workload and issues concerning selective attention. Avoiding excessive information is advantageous not only to elderly but also to people with intellectual disabilities and/or with autism. This can for example make use of symbols. It is also important to keep in mind that some noises and excessive information can be quite disturbing for people with autism. Moreover, it is essential to implement different sensory channels. Visually impaired users, for example, need both audible and/or braille format, and would also benefit from large fonts, contrasting colours and appropriate illumination. These features would likewise serve elderly passengers and those with hearing disabilities that cannot rely only on audible information systems. Additionally, for these passengers, a proper illumination is crucial for lip reading.

The above mentioned mistrust of technology can also be addressed through HMI. An interview study by Li et al. (2019) focuses on the general design of an age-friendly highly automated vehicle. As mentioned, some people, especially elderly, find it difficult to fully relinquish control. They require information and a monitoring system, to be able to control the behaviour and the decision making of the automated vehicle. Additionally, the takeover requests of an automated vehicle should be adjustable and explanatory. The driving style should be imitative and corrective, such that it imitates the standard driving style of the driver and corrects bad and dangerous driving behaviour at the same time.

The main requirements of passengers with physical limitations, like elderly or people with disabilities, such as wheelchair users, are much more oriented towards the main body and internal arrangement of the vehicle that needs to be accessible and barrier-free (i.e., have wide and step-free doors, easy to use door handles). Therefore, as with current accessible vehicles, adjustable floor to accommodate to wheelchairs, or a ramp/ lift system is needed, as well as adequate space for manoeuvring the wheelchair. Furthermore, to ensure a smooth continuation of the passenger's trip, the AVs should navigate to accessible disembarking locations where, for example tactile pavement and audible traffic lights exist, or barrier-free space away from traffic.

This list is not exhaustive and could potentially be extended through consultations with the heterogenic groups of persons with disabilities. But what is clear from the results and examples shown is that the requirements of different vulnerable groups for AVs vary greatly depending on the individual user's impairment.

4.2 Making Use of Universal Design

In 1997, a working group of architects, product designers, engineers and environmental design researchers from the North Carolina State University developed the 7 Principles of Universal Design (NCSU 1997):

1. Equitable Use;
2. Flexibility in Use;
3. Simple and Intuitive Use;
4. Perceptible Information;
5. Tolerance for Error;
6. Low Physical Effort;
7. Size and Space for Approach and Use.

The purpose of the principles is to guide the design of environments, products and communications. According to the Center for Universal Design in NCSU (1997), the principles "may be applied to evaluate existing designs, guide the design process and educate both designers and consumers about the characteristics of more usable products and environments."

The term universal design is often used interchangeably as design-for-all, which includes the respect of human diversity. Instead of removing barriers, the method focuses on prevention. Products, services, systems and the surrounding environment is designed in such a way that the final product of the design is usable and accessible for the widest possible range of people regardless of their age, gender or capabilities. As a result, more users can be reached and production costs can be reduced by sharing them among a larger market. Moreover, products can adapt to the changing needs through our lifetime without costly and burdensome alterations due to their flexibility.

These cost benefits of accessibility have been supported by surveys, e.g. the European Commission published its Impact Assessment accompanying the document Proposal for European Accessibility Act in 2015 (EU-KOM. 2015). Annex 2 of the document contains the results of the Stakeholder consultations where companies were asked to provide information about how accessibility is considered when providing goods and services, and estimates of the costs and benefits of accessible goods and services. The great majority of the 180 respondents were micro, small and medium enterprises. The respondents generally regarded the extra costs of accessibility to be relatively low, at less than 5% of production costs. 55% of companies that provide accessible goods and services have increased their clientele as a result of improving the accessibility of their goods and services, and 39% have experienced increases in their financial benefits for this reason.

The UN CRPD (2006) mentioned in the beginning brought the definition of universal design to an international binding legal regulation. On page 4, Article 2 of the UN CRPD includes the following definition of universal design: "Universal design means the design of products, environments, programmes and services to be usable by all people, to the greatest extent possible, without the need for adaptation or specialized design. Universal design shall not exclude assistive devices for particular groups of persons with disabilities where this is needed."

That means, for any design to be inclusive, the built environment, products, services, information and communication technology should work well for everybody. That relates to the transport infrastructure and vehicles, as well as, the transport service. Accessibility and usability are being shifted from being optional to mandatory and it is based on the universal design for all approach.

As a further step, in 2019, a new CEN standard EN 17161:2019 entitled 'Design for All - Accessibility following a Design for All approach in products, goods and services - Extending the range of users' was published (CEN 2019). The document helps an organisation to meet their statutory and regulatory requirements regarding accessibility of its goods and services. It promotes accessibility following a Design for All approach in mainstream products, goods and services and interoperability of these with assistive technologies. However, this document does not provide technical design specifications and does not imply uniformity in design or functionality of products, goods and services. The standard is a tool to mainstream a universal design approach throughout the internal process of manufacturers and service providers, which would result in more accessible goods and services.

For improving the access of vulnerable groups to AVs and also the interaction of such persons with AVs in road traffic, the application of Universal Design in the development of the corresponding HMI is a key element. However, Dey et al. (2020) found this is where a major gap still exists. Most existing HMIs for the interaction between human road users and AVs use a single modality (i.e. lights or sounds) and would thus not address road users with special needs, such as people with vision or hearing impairments. Same can be stated for HMIs for the interaction between passengers and vehicles.

But all guidelines are of little help as long as the very differentiated needs identified above are not considered in the development. Therefore, products like AVs which are moving on public roads and consequently have to meet the requirements of a wide range of people including such with disabilities cannot be created without close cooperation between designers and end-users. This is where the concept of co-design has to come in. 'Co-design is an approach to the discovery, definition and design of products, services and environment that invites the end-users into the design process as active participants. The co-design approach leverages a combination of methods and tools to gain deep insights about people's experience, latent needs, dreams and aspirations (Sanders 2021).

And furthermore, article 4 (f) UN CRPD states: 'General obligation of the UN CRPD stipulates that States Parties shall undertake or promote research and development of universally designed goods, services, equipment and facilities, which should require the minimum possible adaptation and the least cost to meet the specific needs of a person with disabilities, to promote their availability and use, and to promote universal design in the development of standards and guidelines.'

5   Good Practice Example: HADRIAN Shows a User Centred Development Approach to Enable Diverse Mobility Needs

A practical example of how user needs should be researched in order to shape the development of automated vehicle functions is the HADRIAN project. Within the project, user-centred mobility solutions for AVs were developed for SAE levels 2 and 3, taking into account different users and their individual (mobility) needs.

The HADRIAN project primarily aims to shape automated driving from a holistic, user-centred perspective by addressing three main pillars: First, a fluid Human Machine Interface (f-HMI) helps drivers and users to appropriately interact with the vehicle. Second, through integrating the AV with road infrastructure the AV is made more predictable and available. Thirdly, an onboard tutoring application teaches the driver to develop safe AV usage skills over time and improve the safety. These innovations practically extend the SAE automated driving levels (SAE ADL). Specific mobility scenarios concerning individual user as well as mobility needs and constraints, have been developed. These mobility scenarios are an important basis for the development of system functions, simulations and tests during the HADRIAN projects. Specific personae have been conceived which all will benefit from the automated driving functions in their specific user context. The imaginative description of their specific driving tasks significantly help system developers designing the technical applications in a user-friendly way according to their anticipated needs.

The HADRIAN partners identified elderly drivers as one important user group that can benefit from the fluid HMI functions. Hence, one of the HADRIAN uses cases describes Harold, a 78 year-old man living in the suburbs of Paris, which start to encounter some difficulties driving his car. Based on Harold, three potential mobility scenarios, including potential obstacles on the way (e.g. difficult intersection, highway entry), have been designed and used as a guideline for the development of the specific HADRIAN f-HMI components.

The first scenario describes a visit of his daughter who lives in the countryside, where Harold gets an adaptive information assistant (sensing system), giving situational information about the state of the environment and depending on the driver's Fit-to-Drive (F2D) value. It presents an extension to partial driving automation (SAE ADL 2). Harold always stays in the loop of driving but is supported through an adaptive assistant. Only if safety cannot be ensured, a planned emergency stop is executed.

In a second scenario, Harold goes on vacation to an unknown place at the sea. This is an extension to conditional driving automation (SAE ADL 3). Assisted driving takes over when Harold seems not to be able to complete the driving process and actively suggests transitioning from manual to automated driving level (SAE ADL 3).

Finally, Harold wants to visit his doctor in the city. During this trip he will be provided with a "Guarding Angel Protection". This is an extension of SAE ADLs 3–4 (conditional/ high) driving automation. A Guarding Angel functionality provides a self-enabling safety mechanism against accidents and keeps the vehicle within safe operational boundaries while allowing Harold to actively manoeuver the vehicle within those boundaries. Emphasis is on supporting Harold where necessary without overwhelming him, offering him information or supporting takeover functions as needed and asked for with high transparency of actions (Fig. 1).

a   **b**   **c**

Fig. 1. Three Scenarios of Harold. **a** daughter visit in the countryside, **b** vacation at the sea, **c** doctor visit © 2019 Hadrian.

The above described mobility scenarios led to detailed application descriptions (DAD) as design basis for the development of the specific HADRIAN f-HMI components, including "Increased Automated Driving Predictability", "Human-Centred Fluid HMI", "Adaptive Seating Orientation", "Visual Head-UP Display", "Haptic Feedback on Steering Wheel", "Ambient Light Indicators", "Fluid Interface Design", "Tutoring System". Hereby a task analysis investigated the implications of using the HADRIAN innovations (see Fig. 2) within the mobility scenarios, leading to initial requirements for vehicles, road infrastructure and the driver/user for the HADRIAN operational concept.

Figure 2 also shows which HADRIAN Innovations (indicated with HI# at the top) support the Harold mobility scenarios (indicated with H# at the bottom of the figure). In the first mobility scenario (H1), Harold is supported by an awareness assistant (HI1), active driver monitoring & fluid interventions (HI5) and adaptive tutoring (HI6), marked though the lines connecting the HADRIAN Innovations and the Modes of Automation. Those innovations serve as a manual driving aid for elderlies. In the second mobility scenario (H2) support will be realized by providing minimum guaranteed time for human driver to transition from automated driving to manual driving (HI3), guaranteeing minimum duration of automated driving at level 3/3+before the trip (HI4), as well as HI5 and HI6. During the third mobility scenario (H3) HI5, HI6 and the Guarding Angel safety protector (HI7) support Harold.

The DAD are the basis for legal and ethical considerations for the HADRIAN operational concepts as well as considerations for driver information needs, knowledge and skills. The special needs of elderly drivers and the corresponding DAD have been discussed and verified in focus group discussions with invited elderly people. The participants (65+years) were introduced to Harold, his main driving challenges and four to

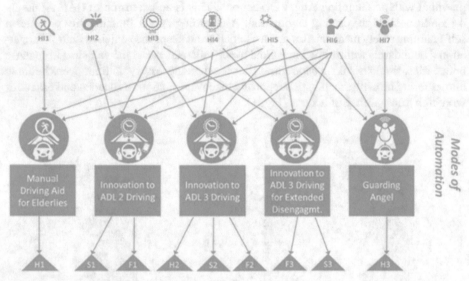

Fig. 2. Relation of main technical and procedural HADRIAN innovations, modes of automation and the corresponding mobility scenarios © 2019 Hadrian.

six scenario segments with specific driving situations. Detailed description and presentation of the scenarios allow a realistic understanding of the situations, which might come up in connection with automated driving. Following the video presentation, they were asked to rank the importance of given ethical values "privacy", "autonomy of the driver", "safety", "security", "vehicle performance" and "costs" in relation to the different driving scenes. This specific user group discussion provided basic information for the HMI design in the vehicle. User-centred design principles could be developed. Additionally, the potential risks concerning the driver vehicle interactions are evaluated with respect to the DADs, and tests and simulations are operated in alignment with the different mobility scenarios.

The identification of mobility needs of specific user groups and the translation into detailed requirements for the vehicles and, in this case, the HMIs is a recommendable approach to the start of the technical development of new functions. Road infrastructure, vehicles and drivers themselves will benefit from greater acceptance within that user group. The automated driving innovations will be relevant to the users and also support inclusivity. Therefore, this approach will lead to a successful implementation of automated driving, also for wider and more versatile user groups.

Specifically, the Harold mobility scenarios provide an example for the use of vehicle automation on levels 2 and 3 for the benefit and inclusion of elderly people. Certain impairments of elderly people and based on that specific mobility needs, mobility challenges and driving requirements are also relevant for other vulnerable user groups (e.g. people with physical disabilities, novice drivers). By designing the "diving process", a more detailed picture of the actual needs for certain user groups can be drawn and user specific HMI components can be develop or adjusted. HADRIAN shows how to approach this with view on different and special user needs. This approach can easily be transferred to other vulnerable user groups. Furthermore, the focus on inclusive solutions for automated driving systems and the HMI will be beneficial for all vulnerable user groups. Ongoing discussions on legal and ethical issues and safety concerns of automated driving can only be profitable.

6 Conclusion

One important function of transport systems is to provide accessible mobility for all. Legal requirements both in the EU and in its member states have grown to address the needs of everybody in new investments and legal frameworks have been established to ensure non-discrimination and equality. Increased life expectancy and ageing of the population requires countries to establish policies, which enable the elderly and people with disabilities to live independently and be active members of society for as long as possible. However, the ongoing transformation of the mobility sector has not always been user-oriented, neglecting at times the needs of different people. To ensure real inclusiveness, it is important to consider the needs of every citizen when launching new commodities in society.

The implementation of AVs forms an opportunity to empower vulnerable citizens, namely people with disabilities and elderly people in fulfilling their mobility needs. This technology development will help people with disabilities and age-related impairments to live more independently, with the additional benefit of improving their participation possibilities in education and the labor market, and in general having an as active life as they want.

On the other hand, like many (digital) technological developments, there are still several hurdles to overcome to fully address the needs of vulnerable groups during the deployment of automated systems and vehicles. Universal design as an essential tool to accessibility must play a major role in the development methods of manufacturers. By following this design method, AVs will be usable and accessible for as many user groups as possible. Therefore, knowledge and good practices of Universal Design shall be promoted among AV and HMI developers from an early stage on.

To ensure that the HMI used in AVs, transportation infrastructure, and the vehicles themselves are usable and accessible to all, potential users, especially people with disabilities, must be meaningfully included in the design process throughout the development process until the final product is presented. Co-creation and participative initiatives are critical methods that offer this possibility, bringing together industry experts and developers, researchers and users to work together and develop better products. This has in fact been demonstrated in projects such as HADRIAN where the user needs have been the main focus of the work and were well integrated into the development process. Members of vulnerable groups have established organizations from grassroots to national and even European and global level. The involvement of these organizations' representatives in the development and design of autonomous vehicles will lead to more accessible and usable solutions.

Acknowledgement. This chapter has been developed in the framework of the European Project "HADRIAN: Holistic Approach for Driver Role Integration and Automation Allocation for European Mobility Needs", funded by the European Commission under the European Union's Horizon 2020 research and innovation program. Grant agreement No 875597.

References

Acheampong, R.A., Cugurullo, F., Gueriau, M., Dusparic, I.: Can autonomous vehicles enable sustainable mobility in future cities? Insights and policy challenges from user preferences over different urban transport options. Cities **112**, 103134 (2021). https://doi.org/10.1016/j.cities

BMVI: Ethik-Kommission Automatisiertes und Vernetztes Fahren - Bericht. German Federal Ministry for Digital an Transport (2017)

CAVForth: Welcome to CAVForth. The world's most ambitiuos and complex autonomous bus pilot (2022). https://www.cavforth.com/. Accessed 30 Aug 2022

CEN: CEN EN 17161:2019. Standard. Design for All - Accessibility following a Design for All approach in products, goods and services - Extending the range of users (2019)

Clery, E., Kiss, Z., Taylor, E., Gill, V.: Disabled people's travel behavior and attitudes to travel. UK Department for Transport (2017). https://assets.publishing.service.gov.uk/government/uploads/system/uploads/attachment_data/file/647703/disabled-peoples-travel-behaviour-and-attitudes-to-travel.pdf. Accessed 21 July 2022

Dey, D., et al.: Taming the eHMI jungle: a classification taxonomy to guide, compare, and assess the design principles of automated vehicles' external human-machine interfaces. Transp. Res. Interdisc. Perspect. **7**, 100174 (2020). https://doi.org/10.1016/j.trip.2020

Diepold, K., Götzl, K., Riener, A., Frison, A.-K.: Automated driving: acceptance and chances for elderly people. In: Proceedings of the 9th International Conference on Automotive User Interfaces and Interactive Vehicular Applications Adjunct - AutomotiveUI 2017, pp, 163–167 (2017). https://doi.org/10.1145/3131726.3131738

Directive 2013/33/EU of the European parliament and of the council of 26 June 2013 laying down standards for the reception of applicants for international protection (recast) (2013)

European Blind Union (EBU): Facts and Figures (2022). https://www.euroblind.org/about-blindness-and-partial-sight/facts-and-figures. Accessed 30 Aug 2022

European Commission. EU-KOM: Commission staff working document accompanying the document Proposal for a Directive of the European Parliament and of the Council on the approximation of the laws, regulations and administrative provisions of the Member States as regards accessibility requirements for products and services. SWD/2015/0264 final - 2015/0278 (COD) (2015)

European Commission. EU-KOM European comparative data on Europe 2020 and persons with disabilities. Brussels (2020). https://doi.org/10.2767/745317

European Commission. EU-KOM: Statistics on Migration and Europe (2022). https://ec.europa.eu/info/strategy/priorities-2019-2024/promoting-our-european-way-life/statistics-migration-europe_en. Accessed 30 Aug 2022

European Network Against Racism (ENAR): Election Analysis: Ethnic Minorities in the new European Parliament 2019–2025 (2019). https://www.enar-eu.org/enar-s-election-analysis-ethnic-minorities-in-the-new-european-parliament-2019/. Accessed 30 Aug 2022

Eckstein, L.: Safety Assurance – Developing and Assessing Automated Driving. In: Proceedings of Automated Vehicles Symposium 2016, San Francisco. AUVSI and TRB (2016)

Eurostat 1: A look at the lives of the elderly in the EU today. Webtool (2022). https://ec.europa.eu/eurostat/cache/infographs/elderly/index.html. Accessed 30 Aug 2022

Eurostat 2: Urban and rural living in the EU (2022). https://ec.europa.eu/eurostat/web/products-eurostat-news/-/edn-20200207-1. Accessed 30 Aug 2022

Eurostat 3: Gender statistics (2022). https://ec.europa.eu/eurostat/statistics-explained/index.php?title=Gender_statistics. Accessed 30 Aug 2022

Eurostat 4: Living conditions in Europe - poverty and social exclusion (2022). https://ec.europa.eu/eurostat/statistics-explained/index.php?title=Living_conditions_in_Europe_-_poverty_and_social_exclusion&oldid=549030. Accessed 30 Aug 2022

Eurostat 5: Reconciliation of work and family life – statistics (2022). https://ec.europa.eu/eurostat/statistics-explained/index.php?title=Reconciliation_of_work_and_family_life_-_statistics. Accessed 30 Aug 2022

Fagnant, D.J., Kockelman, K.: Preparing a nation for autonomous vehicles: opportunities, barriers and policy recommendations. Transp. Res. Part A: Policy Pract. **77**, 167–181 (2015). https://doi.org/10.1016/j.tra.2015.04.003

Földesi, E.: Disabled people spoke about their benefits and requirements on autonomous cars. European Transport and Mobility Forum Mobility4EU (2017). Accessed 22 May 2022. https://medium.com/@mobility4eu/disabled-people-spoke-about-their-benefits-and-requirements-on-autonomous-cars-232cdb73f197

Földes, D., Csiszár, C., Zarkeshev, A.: User expectations towards mobility services based on autonomous vehicle. 8th International Scientific Conference CMDTUR, pp.7–14 (2018). http://real.mtak.hu/id/eprint/90280

GRC: Charter of fundamental rights of the European Union (2012). http://data.europa.eu/eli/treaty/char_2012/oj

HEAT: R&D project Hamburg Electric Autonomous Transportation. Final Report (2021)

Horswill, M.S., Anstey, K.J., Hatherly, C., Wood, J.M., Pachana, N.A.: Older drivers' insight into their hazard perception ability. Accid. Anal. Prev. **43**(6), 2121–2127 (2011). https://doi.org/10.1016/j.aap.2011.05.035

Kedmi-Shahar, E., Delaere, H., Vanobberghen, W., Di Ciommo, F.: Analysis Framework of User Needs, Capabilities, Limitations & Constraints of Digital Mobility Services. Deliverable 1.1 of INDIMO project (2020). https://www.indimoproject.eu/wp-content/uploads/2022/02/INDIMO-D1.1-FINAL_v2.0.pdf

Kim, S., Hong, J.-H., Li, K.A., Forlizzi, J., Dey, A.K.: Route guidance modality for elder driver navigation. In: Kay, J., Lukowicz, P., Tokuda, H., Olivier, P., Krüger, A. (eds.) Pervasive 2012. LNCS, vol. 7319, pp. 179–196. Springer, Heidelberg (2012). https://doi.org/10.1007/978-3-642-31205-2_12

Krail M., et al.: Energie- und Treibhausgaswirkungen des automatisierten und vernetzten Fahrens im Straßenverkehr - Wissenschaftliche Beratung des BMVI zur Mobilitäts- und Kraftstoffstrategie (2019)

Li, S., Blythe, P., Guo, W., Namdeo, A.: Investigation of older drivers' requirements of the human-machine interaction in highly automated vehicles. Transport. Res. F: Traffic Psychol. Behav. **62**, 546–563 (2019). https://doi.org/10.1016/j.trf.2019.02.009

Litman, T.: Autonomous Vehicle Implementation Predictions - Implications for Transport Planning. Victoria Transport Policy Institute (2022). https://www.vtpi.org/avip.pdf. Accessed 22 May 2022

Lui, L., et al.: Computing systems for autonomous driving: state-of-the-art and challenges. IEEE Internet Things J. **8**(8), 6469–6486 (2020). https://doi.org/10.48550/arXiv.2009.14349

Martínez-Díaza, M., Soriguera, F.: Autonomous vehicles: theoretical and practical challenges. Transp. Res. Procedia **33**, 275–282 (2018). https://doi.org/10.1016/j.trpro.2018.10.103

Neumann, P., Reuber, P.: Ökonomische Impulse eines barrierefreien Tourismus für alle. Bundesministerium für Wirtschaft und Arbeit, Dokumentation, 526 (2003). https://www.barrie refrei-brandenburg.de/fileadmin/user_upload/_imported/fileadmin/user_upload/tmb_upload/dokumente/Kurzfassung_Oekonomische_Impulse.pdf. Accessed 21 July 2022

NCSU.: The Center for Universal Design (1997). The Principles of Universal Design, Version 2.0. Raleigh, NC, North Carolina State University (1997)

Penmetsa, P., Kofi, A.E., Wood, D., Wang, T., Jones, S.L.: Perceptions and expectations of autonomous vehicles ' a snapshot of vulnerable road user opinion. Technol. Forecast. Soc. Chang. **143**, 9–13 (2019). https://doi.org/10.1016/j.techfore.2019.02.010

Reuters Waymo self-driving vehicles cover 20 million miles on public roads (2020). https://www.reuters.com/article/us-autonomous-waymo/waymo-self-driving-vehicles-cover-20-million-miles-on-public-roads-idUSKBN1Z61RX. Accessed 22 May 2022

Rojas-Rueda, D., Nieuwenhuijsen, M., Khreis, H.: 1859 – autonomous vehicles and public health: literature review. J. Transp. Health **5**, S13 (2017). https://doi.org/10.1016/j.jth.2017.05.292

Roos, M., Siegmann, M.: Technology Roadmap for autonomous driving. Working Paper Forschungsförderung, Vol.188. Düsseldorf (2020).

SAE International: Taxonomy and Definitions for Terms Related to Driving Automation Systems for On-Road Motor Vehicles. SAE Standard J3016−202104 (2021). https://www.sae.org/sta ndards/content/j3016_202104/

Sanders, L.: A Universal Design for All Approach: Methods for a Culture of Co-Design, Webinar (2021). https://universaldesign.ie/products-services/webinar-a-universal-design-for-all-approach-methods-for-a-culture-of-co-design-/

STIMULATE: R&D project Urban sustainable Mobility using electric automated Minibuses. Final Report (2021)

Tesla: Future of driving. (2022). https://www.tesla.com/autopilot. Accessed 22 May 2022

Mercedes Press Release: Conditionally automated driving: Mercedes-Benz announces sales launch of DRIVE PILOT (2022). https://group-media.mercedes-benz.com/marsMediaSite/en/instance/ko.xhtml?oid=53213668&ls=L2RlL2luc3RhbmNlL2tvLnhododG1sP29pZD05MjY 1NzUwJnJlbElkPTYwODI5JmZyb21PaWQ9OTI2NTc1MCZyZXN1bHRJbmZvVHlwZU lkPTQwNjI2JnZpZXdUeXBlPXRodW1icyZzb3J0RGVmaW5pdGlvbj1QVUJMSVNIRURf QVQtMiZ0aHVtYlNjYWxlSW5kZXg9MSZyb3dDb3VudHNJbmRleD01JmZyb21JbmZZ vVHlwZUlkPTQwNjI4&rs=1. Accessed 22 May 2022

European Parliament: Regulation (EU) 2021/782 on rail passengers' rights and obligations (recast) (2021). https://eur-lex.europa.eu/legal-content/EN/TXT/PDF/?uri=CELEX:32021R0782&fro m=EN

UN CRPD: United Nations Convention on the Rights of Persons with Disabilities (2006). https://www.un.org/development/desa/disabilities/convention-on-the-rights-of-persons-with-disabilities.html. Accessed 05 Sep 2022

UNECE: Proposal for new UN Regulations on uniform provision concerning the approval of vehicles with regards to cyber security and cyber management system (2020).. https://unece.org/DAM/trans/doc/2020/wp29grva/ECE-TRANS-WP29-2020-079-Revised.pdf

UNHCR: Master Glossary of Terms Rev. 1 (2006)

VOLVO: Autonomous driving (2022). https://www.volvotrucks.de/de-de/trucks/alternative-ant riebe/autonome-lkw.html. Retrieved 30 Aug 2022

Vaa, T.: Impairments, diseases, age and their relative risks of accident involvement: results from meta-analysis. Transportøkonomisk institutt (2003)

Weinrich, R.: No more blind spots. DEKRA Solutions (2017). https://www.dekra-solutions.com/2017/12/no-more-blind-spot/?lang=en. Accessed 22 May 2022

Citizen-Centered Mobility Model of Catalonia

Xavier Sanyer Matias[✉] and Lluís Alegre Valls

Metropolitan Transport Authority of Barcelona, Barcelona, Spain
xsanyer@atm.cat

Abstract. The mobility is becoming ever more complex in a changing environment with the emergence of new means of transport and mobility solutions. In this ecosystem in flux, authorities responsible for mobility must plan for the arrival of such changes, considering the desired mobility model to be attained, one where sustainability, health, digitalisation, and equity, among other aspects, take a prominent role; and all this without disregarding the public, who must be placed at the centre. The digitalisation occurring in society today offers a wide range of solutions; these digital solutions must consider that there is a percentage of the population who are non-digital and that its deployment must the different types of territories, urban and rural.

The analysis set out in this article examines the case of the Barcelona Metropolitan Region, describes the characteristics of its citizens from a standpoint of mobility and their acceptance of new trends, identifies existing aims and strategies regarding mobility in the region, and specifically distinguishes the digitalisation strategy in this European region. Finally, this analysis must consider the changes brought about by the COVID-19, in which certain habits have been altered, and where digitalisation has played and will continue to play a sizeable role.

1 Existing Planning Elements

The mobility ecosystem has been undergoing changes in recent times. For example, new factors have emerged, such as shared mobility systems, the concept of mobility-as-a-service, scooters, or the return of the bicycle to the city. In addition, certain pre-existing means of transport, with the aid of digital tools, have developed new functionalities that are becoming new transport solutions for the general public.

This is why good mobility planning elements are required, and also why the authorities responsible for mobility must make good use of them. In the case of the Barcelona Metropolitan Region, since 2008 its transport authority, Autoritat del Transport Metropolità, has periodically drafted a mobility master plan for the territory, a document that lays out the objectives and lines of action in reference to the mobility policy in this region of over 5 million inhabitants. In its most recent version for the 2020–2025 period, the plan encompasses 5 broad two-fold objectives for the planned mobility model, which are described below with more specific goals within each objective.

I. Keseru and A. Randhahn (Eds.): *Towards User-Centric Transport in Europe 3*, LNMOB, pp. 42–55, 2023.
https://doi.org/10.1007/978-3-031-26155-8_3

Sustainable and Healthy Mobility

- Shift towards more sustainable modes of transport, keeping the distance of journeys to a minimum
- Lower energy consumption and less impact of mobility on climate change
- Improved public health and minimisation of social costs
- Encouraging the public's physical activity

Efficient and Productive Mobility

- Increased efficiency of the transport model, fostering the socio-economic optimisation of the system
- New jobs with particular emphasis on new technology sectors
- Fostering of new business models that leverage opportunities emerging from the circular and innovative economy

Safe and Reliable Mobility

- Reduced accident rates and improved perception of safety
- Reliable public transport system responsible to its users
- Promotion of safe, quality spaces for active modes

Inclusive and Egalitarian Mobility

- Total accessibility of the mobility system
- A mobility system that meets the different needs of the entire citizenry
- Incorporation of a gender- and age-based perspective across the entire mobility system

Smart and Digital Mobility

- Bringing new mobility technology to the general public and business community
- Boosting a digital mobility that services the mobility needs of the public at large
- Readying the mobility system for the challenges brought on by mobility automation

This set of goals is to be met through the realisation of around one hundred measures, divided into 10 pillars of action. The interrelationships between these objectives and measures are shown below (Fig. 1):

Of all the actions contained in the Mobility Master Plan, digitalisation and the actions intended for the public at large play a highly significant role. On the one hand, digitalisation is a widely expanding tool that can optimise and facilitate the development of many of the above actions.

On the other, the solutions intended for the general public are considered essential to achieving the success of the objectives set. Similarly, putting citizenship centre stage is also a key element in achieving the objectives of inclusiveness and equity that have been set. These objectives will be achieved by implementing initiatives to guarantee physical and digital accessibility to the system and designing solutions to meet citizens'

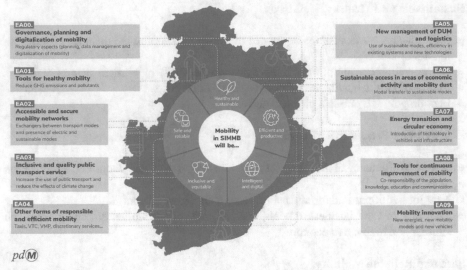

Fig. 1. Objectives and measures of the Mobility Master Plan 2020–2025 for the Metropolitan Region of Barcelona

needs, such as the appropriate charges and digital solutions providing solutions to real problems, and conceiving actions for vulnerable collectives by facilitating access and training in the use of digital tools.

2 Citizen Mobility and Attitude Towards Changes

Broad knowledge of the public's mobility habits and characteristics are important to developing the right mobility policy, with the aim that these solutions are in their interests. This is why since 2003 the planning authorities of the Barcelona Metropolitan Region have regularly performed a weekday mobility survey of more than 10,000 residents across the region, which makes a detailed analysis of their mobility, the evaluation of the different mobility solutions and certain aspects related to future changes.

The latest data available in the 2021 survey report active mobility to be by far the most used form of mobility in the Barcelona Metropolitan Region, as it accounts for 47.9% of the total number of trips. This is followed by the data from private vehicles, which has a share of 37.1% of all journeys; and finally, public transport, with a 15% share (Table 1).

However, new actors have emerged on this mobility scheme, of little importance in absolute numbers at the moment, but with sizeable growth in recent years. To give an example, in 2017, the number of trips in the Barcelona Metropolitan Regions that were made using a scooter or equivalent systems is estimated to be 13,000, while the figure for the same category for 2021 is 110,000 trips, or in other words, 9 times higher in just 4 years.

In the case of this explosive growth, the consequences have included a significant increase in the number of disputes with pedestrians because, although they still only

Table 1. No. of trips and modal split in the Metropolitan Region of Barcelona (Metropolitan Transport Authority of Barcelona, 2021). https://www.omc.cat/en/w/working-day-mobility-surveys-emef-

Mode of transport	No. of trips	%
Walking	8,214,153	45.5%
Cycling	308,482	1.7%
Wheelchair	14,403	0.1%
Scooter or equivalent	110,402	0.6%
Active Mobility - Total	8,647,440	47.9%
Bus	998,527	5.5%
Metro	931,732	5.2%
Rail	649,003	3.6%
Other Public Transport	125,565	0.7%
Public Transport - Total	2,704,826	15.0%
Car	5,707,550	31.6%
Motorbike	729,357	4.0%
Van, truck and other private	264,400	1.5%
Private Vehicles - Total	6,701,307	37.1%
Total	18,053,573	100%

represent under 1% of the total number of trips, they happen to be concentrated in certain parts of the city and also interact with pedestrians on the pavement, where areas of tension are generated and mounting problems. It must be taken into account that many of these new forms of mobility are based on digital solutions.

Consequently, as the above data indicate, we are faced with very rapid changes, ones to which the planning authorities must react in order to correctly introduce these systems in the urban and metropolitan ecosystems, while also taking into consideration the acceptance and interest of the end users.

This aspect has also been analysed in the Metropolitan Region in recent years, with the support of complementary studies which have explored the public's acceptance of the changes and solutions put forth in this evolving mobility ecosystem. Knowledge of the receptivity to these changes and solutions by society at large is considered essential to ensure that planning authorities can anticipate the public's mobility patterns in the future.

The most recent edition of one of these complementary studies is from 2019, which provides some relevant data to be able to identify the public's interests and habits and future mobility trends, the majority of which are linked to the digitalisation of mobility. The data collected are based on a total of 3,000 interviews with citizens of the Barcelona Metropolitan Region.

Conducting these types of studies can help identify different types of citizens according to their mobility habits and also evaluate aspects such as the improvement of public transport, the implementation of low-emission zones, the use of digital tools, the introduction of new means of transport, as well as aspects of new consumer habits, such as online commerce, among other aspects.

For the purposes of this article, some of the most significant aspects of digital mobility solutions have been identified in the aforementioned studies, as well as the public's opinion of them within the consideration of a user-centric digital solution. The study data show a sizeable increase in users who use tools to keep informed about the functioning of the different transport options, rising from 32% in the 2017 edition to over 50% in that of 2019. This increase is basically due to purely digital tools, specifically in those controlled by large international corporations, such as Google or Waze (Table 2).

Table 2. Information channels most used by citizens to plan their trips in the Metropolitan Region of Barcelona (Metropolitan Transport Authority of Barcelona, 2021).

Information channel	% (2019)	Information channel	% (2017)
Google Maps	58.4%	Google Maps	55.2%
Other Local Apps	32.5%	Other Local Apps	38.9%
TV-Radio	24.1%	TV-Radio	32.5%
Social Network	12.7%	Social Network	14.5%
Waze	12.3%	Waze	8.2%
Mobility websites	9.3%	Apple Maps	5.1%
Apple Maps	5.7%	Moovit	4.6%
Moovit	4.7%	Wazypark	1.7%
Mou-te (ATM Barcelona & Government of Catalonia app)	3.0%	Others	0.3%
Smou	1.7%		
Others	0.9%		

Analysing the data from the same study, there is also evidence of positive changes with regard to the interest of the public in mobility solutions such as Mobility as a Service and the use of sharing systems, both based on digital tools. In this regard, the use of shared mobility services rose from 30% in 2017 to 40% in 2019, while the concept of Mobility as a Service, which already had a high level of interest in 2017 among 57% of users, increased to 61% in 2019.

Finally, it is worth noting that the study explored the public's opinion with regard to the use of the data generated by these digital tools in order to have available information on citizens' mobility habits. When they were asked explicitly about this issue, more than 70% accepted that their data were used, provided they were appropriately processed, while 48% expressed the view that it is important that such collection and data analysis should be performed by a public body, while 38.3% said it should be done by a public-private partnership. The studies and surveys that provide insight into the mobility and attitudes of the general public in light of these changes are not the only source of information, however, as it is precisely the digital tools themselves which have become complementary elements in characterising mobility.

In 2017, the Barcelona Metropolitan Transport Authority created the first mobility matrices that describe the flow of people and goods using mobile telephone data, a source of information has been consolidated in recent years with the monitoring of mobility flows through individuals' mobile phones. The creation of mobility matrices as a planning element presents advantages and drawbacks with respect to the use of surveys. On one hand, it is worth mentioning the positives of being able to identify a greater number of movements, which provides for greater capillarity; however, more qualitative information and demand segmentation data are lost, as is what the purpose of the trip might be or the modal distribution. That said, telephone data provides for more immediate availability as compared to survey data, so during the COVID-19 pandemic, for example, it was a useful source of data for many planning entities to track mobility and its evolution, as the Barcelona Metropolitan Transport Authority did.

3   The Impact of COVID-19 on Certain Mobility Habits

As mentioned at the beginning of the article, COVID-19 has been a disruptive factor causing changes in mobility, socialisation and consumption habits, and even today we do not know for sure if such changes will persist, structurally-speaking. Triggered by the public health pandemic, these changes have largely taken place due to the presence of digital solutions, whether to support the changes in people's habits or in the use of new means of transport. From the start of the pandemic, the Barcelona Metropolitan Transport Authority has taken an interest in being aware of these changes and forecasts of transport use by the public at large as well as the impact of COVID-19. To do so, it has carried out several studies both amongst public transport users and the region's business community in search of insights into the changes expected in mobility as a result of the new situation generated by the COVID-19 pandemic.

The first of these studies was conducted by the Mobility Observatory of Catalonia (2020a) during the spring and summer of 2020, which identified the changes in the habits of transport users. The second study of Mobility Observatory of Catalonia (2020b) analysed aspects of the changes in habits linked to work commuting from a corporate standpoint.

The study of the change of habits of transport users showed a predisposition to change in the months that followed the onset of the COVID-19 pandemic. Users showed concern about the use of collective transport systems, at the same time as they took a positive view of individual means of transport, both mechanised varieties and those of active mobility.

When asked about the degree of confidence generated by the different means of transport, underground railway was the one that reflected the lowest degree of confidence, while bicycles and private vehicles were those that had the highest level. The initial predictions for a modal change foresaw decreases in the use of public transport of up to 20% with respect to the number of users prior to the pandemic. These figures, in the end, seem not to have materialised, as in the case of the Barcelona Metropolitan Region, the percentage of public transport use stands at 90% of the pre-pandemic figure.

The study of the business community revealed a fairly high willingness to implement teleworking as a means of reducing the number of trips, but a more lukewarm response to other solutions, such as flexible schedules. The key to the success of the sudden and widespread implementation of teleworking was the use of established digital tools, which managed to maintain productivity levels without the need to travel.

The initial results reflected a strong interest in teleworking, but this has subsequently receded, stabilising in values which, in the case of the Barcelona Metropolitan Region, stand around 13% of the population who perform some kind of teleworking to a greater or lesser extent. These opinion surveys have been complemented by mobile telephone data analysis, which has provided for the monitoring of the changes in citizens' mobility habits, noted in the previous section.

Thus, the changes in the mobility ecosystem, the knowledge of the general public's willingness to accept new digital tools, and the acceleration of the changes as a result of COVID-19, have forced authorities to draw up a road map of digitalisation within their planning elements, as it has become a fundamental tool in the management and range of mobility solutions on offer.

4   The Mobility Digitalisation Strategy

This is why since 2020 there have been initiatives to define the mobility digitalisation strategy in the Barcelona Metropolitan Region. Within the framework of Catalonia as a whole, the Mobility Digitalisation Agenda in Catalonia has been put forward. The Mobility Digitalisation Agenda in Catalonia (ADMC) is a document whose aim is to provide a strategic vision of how to implement the digitalisation process for mobility in Catalonia over a 10-year period (2020–2030), so it is useful to identify relevant aspects from it that will have an impact on the future of mobility in this European metropolis.

The aim of the agenda is to offer an overview of the various challenges related to the digitalisation of mobility in Catalonia, and how these challenges may be related and prioritised. The contents of this agenda are set out in a number of overall objectives, resulting in 7 lines of actions that encompass 26 measures. Taking into account the speed of the changes occurring in the digital field, documents of this kind must be understood as a living document; this means their contents must be reviewed periodically to ensure their relevance and the validity of the mobility digitalisation strategy with regard to the new challenges posed by the emergence of new solutions and technologies throughout its time horizon.

According to the reflections contained in the Agenda and considered relevant in terms of designing a user-centric digital mobility system, the following objectives must be encompassed. Firstly, to lead a digital transformation of the mobility system as a means of moving towards a more sustainable, productive, efficient, inclusive and digital mobility model.

On the other hand, digital technologies must be fostered in mobility in order to offer users greater efficiency and a better experience, and enhance the capabilities and competitiveness of service operators and mobility infrastructures. Furthermore, the objectives must include greater availability for mobility operators to the necessary resources and factors (ICT infrastructure, regulation, knowledge and training in digital technologies, data, energy needs, etc.) to encourage the digitalisation of their services, assets and organisational models. During the digitalisation process, a digital mobility system must be developed in a manner that guarantees the security and privacy of information, system interoperability, business competence and universal access for the society at large and throughout the region.

In terms of promoting digitalisation, there must be an adaptation of the existing business and industrial fabric within the new mobility network, encouraging both the attraction and development of new innovators and sector leaders in digital technology, steering them towards the improvement of mobility services, infrastructure and teams.

In this context of mobility digitalisation, it is essential to further the business community's global competitiveness in mobility and logistics services. Digitalisation must consider the individual as its core element, which is why education, information and awareness-raising must be fostered among the general public in the use of the new digital mobility models, so as to leave no one behind and with the aim that these solutions be used by the greatest number of people.

In order to achieve this objective, the necessary complementary actions must be implemented to ensure that potentially vulnerable users are not excluded. Thus, in addition to the actions on the agenda, which must design solutions encompassing vulnerable collectives and guaranteeing accessibility in digital terms, parallel work should be carried out with other governmental departments to identify collectives which, despite the simple and user-friendly design, may be excluded. These tasks should be designed to include as many users as possible and they should be coordinated with those responsible for education (children), social welfare (elderly people) and immigration (newly-arrived citizens). Finally, a governance model must be promoted for mobility digitalisation based on the coordination between the mobility authorities and digital policies, as well as collaboration with and between private sector entities.

The Agenda considers this feasible to achieve these objectives, provided a set of actions is developed encompassed by the following lines of action:

- Data management and modelling
- Infrastructure to enable digital transformation
- Digital mobility planning and management
- User-centric mobility
- A logistics system based on new technologies
- Participation of the business community in leading the digitalisation
- Managing the change and the digital transformation

As can be seen, these 7 lines of action include one based on user-centric mobility, with the understanding that its basic goal is that mobility should be digital and user-focused, as we have attempted to develop from the beginning of this article.

An analysis of these lines of action provides an insight of the trends currently at work within the sphere of mobility digitalisation. The first proposes the management and modelling of data, as it is important to design a system that integrates information from the mobility systems in digital format as the foundation for the future mobility management and its advanced planning.

This system must guarantee data standardisation, a shared use and both security and privacy. To do so, certain aspects must be developed, such as the data necessary to define the mobility system, greater insight in terms of gender, vulnerable sectors, etc.; standardise the format and process of data collection: and at the same time integrate mobility data into a common point of access to facilitate planning, management and the innovation of new services: Finally, there must the assurance of data security and privacy. There is a clear need to identify different types of citizens while also guaranteeing the security and privacy of their data, which are fundamental to establish a user-centric solution which considers their interests and protects their rights, as indicated in the analysis of the studies conducted in previous sections.

The second line of action, infrastructure to enable the digital transformation, seeks to identify and develop the infrastructures that will provide for the digital transformation of mobility, by means of data collection and guaranteed coverage of the wireless connection. The developments proposed in this second line of action include the following aspects, such as the SMART infrastructures: Sustainable, Multifunctional, Automatic, Resilient and Technified; having the necessary infrastructure for data capture, and finally guaranteeing standardised connectivity throughout the territory to support digital mobility applications.

The third of these lines of action proposes digital mobility planning and management. With regard to this objective, it foresees the creation of a digital inventory of the current transport infrastructure network as the basis for optimising the planning and management of mobility services and supporting an agile, coordinated and data-based decision-making system. Specifically, it proposes to develop aspects such as improving the planning of mobility options and including the requirements of minority groups in decision-making (gender, age, etc.), a coordination of efforts and insights from the authorities responsible for mobility planning and management, the availability of advanced analytical tools to support evidence-based decision-making, optimising the use of the public space with agile, real-time responses, as well as a digital transformation as the facilitator of environmental and public health goals. As it can be seen in this third line of action, once again there appears the need to include the requirements of minority groups in decision-making, and thus consider user-centric solutions that include all groups of citizens.

The fourth line of action is fully linked to the subject of this article, as it advocates a mobility model that is centred on the individual. It sets out to fulfil this aim through fostering **new digital services** to enhance the **user experience** and **personalise** information, improve support services and offer flexible payments. This is why it specifically foresees the enhancement of new mobility services (new modes of transport), offers new customised products (MaaS - Mobility as a Service, route planners, etc.) and improves the experience of the mobility user while making the payment system more flexible. Thus, the introduction of new personalised services that enhance the experience is a significant aspect to consider in a user-centric digital model. As seen at the beginning of the article, the general public shows a strong willingness to adopt new means of transport and in some cases, these now-introduced means are growing exponentially.

The fifth line of action proposes a logistics system based on new technologies and empowering their use in the area of logistics to meet the new challenges faced by the sector and ensure greater efficiency in the transport and distribution of goods, including last-mile distribution. One might think this lies outside the public interest, despite the rise in online shopping in recent years, also noted in the studies on changing habits previously mentioned in this article; but the measures included in this line of action will surely have a greater impact on the public interest than what one might initially expect.

The sixth line of action proposes the participation of the business community in driving digitalisation forward. In this case, it seeks the support and promotion of the business and research communities of the mobility and ICT industries to develop knowledge and business opportunities in the digitalisation of mobility and the understanding of its effects on the environment. It specifically sets out to stimulate the creation of the research and entrepreneurial fabric to advance digitalisation within an appropriate legislative framework.

The final line of action promotes the management of change and digital transformation, and to do so, a cross-sectional change management model must be defined (taking due account of users, transport workers, private companies and government bodies) to ensure the deployment and operation of the new digital processes, training the agents involved and informing and raising awareness of users regarding the new mobility. This strategy sets out to facilitate the transition towards the digitalisation of mobility, helping vulnerable groups in coping with the changes, raising awareness among users about the use of the data and application of the new technologies, and providing information about the improvements and new transport services/methods.

Thus, the agenda addresses various issues that concern the public according to the studies mentioned earlier in the article. The awareness of the use of data and their proper management will be of great consequence in the deployment of digital mobility solutions.

5   A User-Centric Mobility Model

As can be seen in the agenda's lines of action, users' needs and some of the concerns or changes they expect are present throughout it. A response must be given to these needs so that the public finds the services they require, in a streamlined and straightforward manner, they are offered optimal routes for the trips they take, have full information, including with regard to incidences, so that they receive the support they need in any circumstance (Fig. 2).

In the agenda, these requirements are specifically found in line of action 4, as this is the one devoted to a user-centric digital mobility model. It is proposed to implement this through 3 actions, which will be explored below, as these are the ones that can be most effective in meeting the needs of the general public.

5.1   New Mobility Products, Personalised for the User

The customisation of mobility products requires greater wireless connectivity in public transport and the incorporation of added-value products, so users may enjoy a better travel experience. Examples of this are the on-route information and entertainment systems,

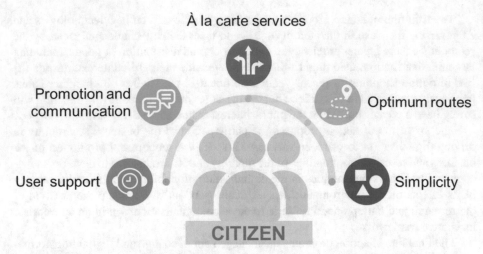

Fig. 2. Characteristics of the services in a user-centric mobility model

MaaS services, and hyper-personalised trip planning tools, which offer the best mobility solutions at any given moment depending on the traveller's needs.

These must serve to improve the user experience when they use public mobility services, in an integrated manner. In order to implement these measures, partnerships should be considered with telecommunication infrastructure operators and mobile phone carriers to study the viability of offering free/discounted Internet access on public transport (Fig. 3).

Finally, it is worth considering that product personalisation can be done by either the public or private sector, so it is important to weigh the option of opening the tender of mobility services dependent on the public administration (including the ticketing system) so that third parties can analyse and make combinations to create new products (custom route planners, MaaS, etc.).

These measures must be reviewed periodically in accordance with the emergence of new technologies that enable new products to be developed for mobility users who have new requirements, which is why it is important to have the involvement of the private sector, as it is known for its great dynamism and a true need for improvement.

5.2   Implementing a Digital Payment System

Digitalisation is a basic tool to enable payment systems on different means of transport. Digital solutions can customise fares to the public's needs and also introduce MaaS. In addition, the implementation of digital payment accelerates and facilitates the creation of new mobility products that are tailored to the user's needs. At the same time, it can generate a large amount of data that provide insight into the public's movements and habits and thus lead to better planning and management of their mobility (Fig. 4).

Digital payment systems have spread around the world in recent years, albeit in coexistence with traditional systems; but each year an increasing number of cities adopt

Fig. 3. Passengers in "Estació de França" station in Barcelona

Fig. 4. A citizen entering public transport with his mobile phone in the commuter trains network in Barcelona

these types of digital solutions, proving that the era of digital solutions has come to stay, with ever-improving features.

5.3 New Mobility Services

Finally, the third of the measures included in this line of action provides for greater availability of information on the supply and demand of mobility systems in order to generate new services and/or business models by third parties (flexible, on-demand services, etc.). This accessibility has to provide for the emergence of new and innovative mobility services that add value to users. This is why it is proposed to open up the information with regard to both supply and demand.

In terms of supply, it is suggested to disclose the information on tenders for mobility services dependent on the public administration (including the ticketing system) so that third parties can analyse and make combinations to create new services. As far as demand is concerned, it is suggested to disclose the information from users (surveys, validations) so that third parties can analyse it and propose new mobility services (new shared mobility services, flexible on-demand services, etc.). This disclosure of the data that new mobility services must allow must further ensure the requirements the public demands in terms of data protection and that their interests are safeguarded.

6 Conclusions

As can be seen, people's mobility is in a process of change due to the transformation our society is undergoing, a process that only accelerated with the pandemic. Digitalisation is an indispensable tool to deal with these changes, one that is in constant development and requires close and constant monitoring by the relevant stakeholders in the sphere of mobility. The public shows a readiness to use new mobility solutions, where the digital factor is of great importance. At the same time, new means of moving around or planning for trips have become commonplace, causing sizeable changes and an increased presence in society at large.

This is why it is important to monitor the public's behaviour and habits, to gain insights of their mobility practices and changes. This monitoring can combine various types of tools, but always with the same aim: to find a suitable solution for the general public, and that in light of how the market is developing, digitalisation plays a crucial part. However, monitoring and analysis of mobility habits is not enough to find the right solution; there must be a strategy to find a sustainable, healthy model, where digitalisation and the public will play a prominent role. It is considered that experiences such as the one implemented in Catalonia may be of interest to other regions, as in a globalised, hyperconnected world, citizens' behaviour is becoming increasingly similar. The driving force of an agenda for the digitalisation of mobility can help tap into certain very broad objectives that not only place the focus on individuals but also the transport of goods, and address issues of infrastructure, management and governance. Among the most significant aspects of this agenda, the ones we believe will attain a user-centric digital mobility model and take into account society as a whole, are the personalisation of solutions, a responsible use of data, and the need for public-private partnership to find the most suitable products. The coming years are likely to evolve rapidly in terms of mobility solutions and the habits of the general public, so it is important for all stakeholders to work in the same direction in their reliance on the digital tools available and consider user-centric solutions.

References

Metropolitan Transport Authority of Barcelona. Working day mobility survey (EMEF) of the Metropolitan Transport Authority of the Barcelona area 2021. https://www.omc.cat/en/w/wor king-day-mobility-surveys-emef-

Metropolitan Transport Authority of Barcelona. Mobility monitoring indicators through mobile phone data (2021). https://app.powerbi.com/view?r=eyJrIjoiZTk5MGJmN2YtNjU0MC0 0NGFjLTk0NTAtMjc1N2EyNTA1NWMyIiwidCI6IjM4ZmY5NzM4LWMwMjgtNGYzNi 1hNjNhLTgxN2I4YWJhNGUwYyIsImMiOjh9

Metropolitan Transport Authority of Barcelona. Analysis of the future behavior of public transport users in Barcelona after the COVID'19 crisis (2020). https://www.omc.cat/documents/662 112/777531/Estudi_comportament_usuaris_Informe_final_v6_Onada.pdf/aac13ca0-49e1-f6fb-e9df-f5c26736710d?t=1639566752657

Metropolitan Transport Authority of Barcelona. Company actions in the field of labour Mobility (2020). https://omc.cat/documents/662112/777531/Actuacions_empreses_mobilitat_lab oral_Informe_Complert.pdf/ad0b2a95-77df-b8ea-b527-b81a9a44dcae?t=1639566743524

Mobility Observatory of Catalonia 2020a. Analysis of the future behavior of public transport users in Barcelona after the COVID'19 crisis (2020a). https://www.omc.cat/en/w/analysis-of-the-fut ure-behavior-of-public-transport-users-in-barcelona-after-the-covid-19-crisis

Mobility Observatory of Catalonia 2020b. Company actions in the field of labour mobility, Final Report (2020b). https://www.omc.cat/en/w/relationship-between-mobility-and-income-of-the-population-in-the-metropolitan-area-of-barcelona-and-the-city-of-barcelona-in-the-con text-of-the-covid-epidemic19-duplicate-1

Who is Left Out: Why Vulnerable People Cannot Access Digital Mobility Services?

Digital Shared Mobility Services: Operationalizing the Capabilities Approach to Appraise Inclusivity

Lluis Martinez[1]([⊠]) and Imre Keseru[2]

[1] Mobilise Mobility and Logistics Research Group, Department Business Technology and Operations (BUTO), Vrije Universiteit Brussel, Brussels, Belgium
Lluis.Martinez.Ramirez@vub.be
[2] Mobilise Mobility and Logistics Research Group, Vrije Universiteit Brussel, Brussels, Belgium

Abstract. Digitalization has fostered the emergence and transformation of transport services, such as shared transport. Digital literacy and having access to digital platforms are increasingly necessary prerequisites to be mobile and benefit from these services. Consequently, new forms of transport disadvantages have emerged, which might result in the exclusion of vulnerable populations.

This paper reviews the literature about transport disadvantages, digital exclusion and shared transport to identify a comprehensive approach to the study of digital shared mobility services (DSMS). By incorporating the digital divide into the Capabilities Approach, a theoretical framework to study DSMS is proposed.

The findings of this paper are relevant to decision-makers, practitioners and researchers working within the field of urban mobility and shared transport services. The theoretical framework proposed is useful to understand the unequal use of DSMS and appraise their inclusivity. This framework is also useful for transport operators and policy-makers interested in adopting a user-centred perspective.

1 Introduction

Digitalization is defined by Gray and Rumpe (2015) as the process in which a wide range of information and communication technologies (ICTs), also referred to as digital technologies, are integrated into all aspects of daily life. During the past decade, digitalisation has accelerated, having a transformative impact on mobility and transport systems (Macharis and Geurs 2019). Citizens increasingly need digital technologies for conducting tasks related to their mobility (Snellen and de Hollander 2017) such as checking schedules, acknowledging incidents, purchasing tickets or booking transport services (Durand et al. 2021). Transport operators have adopted digital technologies as a means to increase cost-efficiency and improve user experience (Davidsson et al. 2016). Moreover, such technologies are the main drivers behind the emergence and development of new transport solutions such as autonomous vehicles, Mobility-as-a-Service (MaaS) and shared transport (Macharis and Geurs 2019; Pangbourne et al. 2020; Shibayama and Emberger 2020).

© The Author(s) 2023
I. Keseru and A. Randhahn (Eds.): *Towards User-Centric Transport in Europe 3*, LNMOB, pp. 59–73, 2023.
https://doi.org/10.1007/978-3-031-26155-8_4

Shared transport is defined as the services that allow users to have short-term access to a transportation mode, such as a vehicle or a bicycle, which is shared with other users (Shaheen and Cohen 2018). Shared transport has become increasingly relevant in urban policy agendas, as a means to potentially reduce congestion levels and greenhouse gas emissions in cities (Cohen and Kietzmann 2014; Machado et al. 2018; Santos 2018). Some scholars even argue that we are currently in an era of shared transport services due to the fast development of solutions and tools that enable the rapid adoption of such services (Shaheen et al. 2016).

Shared transport is highly dependent on digital technologies, with most providers relying on digital platforms to operate their services (Jittrapirom et al. 2017). This requires travellers to have access to a reliable internet connection and a digital device (Groth 2019; Pangbourne et al. 2020). Consequently, not being able or willing to adopt digital technologies may result in a form of transport disadvantage (Schwanen et al. 2015).

Although transport services based on digital technologies, such as digital shared transport, might be especially useful for groups facing transport disadvantages, some of these groups are also at higher risk of digital exclusion (Goodman-Deane et al. 2021). When faced by vulnerable populations, transport disadvantage might result in transport-related social exclusion (TRSE) (Yigitcanlar et al. 2018). In this regard, vulnerable populations are defined as those social groups that suffer from transport disadvantages as a result of their personal characteristics (Maffii and Bosetti 2020). Lucas (2019) refers to TRSE as the form of social exclusion resulting from scarce access to transport services and limited mobility, preventing individuals from reaching necessary destinations and participating in the social life of their community. In this paper, shared transport is not considered a goal in itself, but a means to enable individuals to fulfil their needs more sustainably while reducing transport disadvantages and TRSE. Thus, the study of digital shared transport from a user-centric perspective is considered relevant to enable a transition towards more sustainable and inclusive transport systems.

In current literature, a well-defined framework for the study of DSMS that considers related transport disadvantages and potential forms of exclusion is missing. This results in the lack of a comprehensive understanding of the inclusivity of DSMS. Therefore, this paper aims to fulfil this knowledge gap by proposing a new framework to apprise such services. The following section reviews the literature on transport disadvantages and identifies what population groups are more vulnerable to facing disadvantages when using DSMS. The second section identifies an approach that incorporates the factors that produce such disadvantages and considers the needs of vulnerable groups. Consequently, the Capabilities Approach (Nussbaum and Sen 1993; Sen 1979, 2005, 2009) is adapted to the study of DSMS resulting in a specific framework. To conclude, the last section summarizes the different arguments contained in this paper, proposing further and future advancements.

2   Existing Perspectives on Transport Disadvantages

This paper aims at developing a theoretical framework to allow a comprehensive understanding of the barriers and difficulties that citizens may encounter when using DSMS.

This inquiry starts with a review of existing perspectives on transport disadvantages to compare existing approaches and inform the development of the theoretical framework.

In recent years, transport disadvantages have increasingly been studied from different perspectives, (Jeekel 2018; Pereira et al. 2017). Currie et al. (2010), for instance, define transport disadvantages as the difficulty to reach necessary destinations. Vecchio and Martens (2021), on the other hand, focus on the difficulties to gain accessibility, which they understand as the potential mobility to reach spatially distributed opportunities. Other authors have broadened the understanding of this concept by including the lack of influence on transport-related policies (Hodgson and Turner 2003), or the exposure to negative external impacts, such as pollution or accidents (Feitelson 2002; Schwanen et al. 2015).

Transport disadvantages are a multidimensional construct, as they are the result of the complex interactions between transport systems, land use patterns and individual circumstances (Delbosc and Currie 2011; Jeekel 2018; Páez et al. 2012). This research field can therefore be considered inherently interdisciplinary, resulting in diverging definitions depending on the set of contributing factors considered by the authors. The terminology used in the literature on transport disadvantages includes concepts such as transport poverty, transport justice and transport equity. Thus, the terms transport and mobility services are used in this paper to refer to those services that allow citizens to be mobile, being transported by someone else, as in the case of public transport, or by themselves, as in the case of shared bicycles.

Although such broad terminology might cause conceptual inconsistency (Dodson et al. 2004), in all cases it refers to the distribution of benefits and burdens derived from transportation systems, incorporating central concepts in the transport disadvantages debate, such as equity and justice. The idea of equity is especially relevant for scholars studying the distribution of transport services and related resources (Benenson et al. 2011; Meijers et al. 2012). Likewise, authors that use the term transport justice, also consider equity as the most important criterion. In this case, the concept is used to refer to equal accessibility levels (Martens et al. 2014). Martens et al. (2019, p. 13) define equity as 'the morally proper distribution of benefits and burdens over members of society', while Anderson et al. (2017, p. 65) suggest the following definition: 'ensuring that residents can reach destinations across the city in a time and cost-effective manner, irrespective of their geographic location or socioeconomic status.

The distribution of benefits and burdens derived from transport systems is studied in existing literature from several perspectives. Martens et al. (2014) differentiate three normative approaches which can be found in other scholarly work (Lewis et al. 2021; Pereira and Karner 2021): the egalitarian, sufficientarian and prioritarian approaches. Egalitarianism focuses on the distribution among geographical areas or social groups (Benenson et al. 2011; te Boveldt et al. 2020; Meijers et al. 2012). This approach advocates that everyone should benefit from the same level of services and accessibility and investigates why certain groups or regions have a higher level of accessibility or enjoy better services (Pereira et al. 2017). Sufficientarianism focuses on basic needs, referring to a minimum level of transport services, goods and accessibility that should be available to everybody (Delbosc and Currie 2011). Herein, absolute levels are more important than relative inequalities, all the while highlighting the need for a minimum level of accessibility (Pereira and Karner 2021). It also introduces the idea of transport poverty, referring to the situation of individuals and groups who do not benefit from the minimum acceptable level of transport services (Martens et al. 2014; Pereira et al. 2017). Finally, prioritarianism focuses on the benefits concerning accessibility, advocating increasing benefits for those who suffer more from transport disadvantages (Casal 2007). This perspective combines elements from the two previous approaches, aiming to overcome transport poverty by reducing inequality without necessarily targeting equality (Martens et al. 2014).

3   Factors that Prevent Vulnerable Groups from Using DSMS

Transport disadvantages are experienced unevenly by individuals depending on their characteristics. Populations that encounter a greater number of disadvantages to using a transport service, and as a consequence suffer from low levels of accessibility, are more vulnerable to social exclusion (Jeekel 2018; Lucas 2012; Lucas et al. 2016). Considering how transport disadvantages are experienced depending on the characteristics of an individual is a central step to improving the level of transport services and accessibility of these groups. To allow individuals to better reach necessary destinations and gain mobility, the disadvantages encountered by each individual when using DSMS must be thoroughly considered. In this section, previous research about forms of disadvantages and factors that lead to exclusion are reviewed as a means to identify the vulnerable groups that encounter difficulties to use DSMS.

Although it is widely accepted that improving accessibility is needed to enhance the freedom of choice and equality of opportunities, new perspectives imply that focusing solely on accessibility may lead to overlooking the needs of vulnerable populations (Kuttler and Moraglio 2020). Scholars such as Sheller (2018), argue that increasing accessibility will not improve the mobility of vulnerable groups if the social processes that produce transport disadvantages are ignored. Furthermore, focusing on resources can be misleading, as the needs and abilities of people are heterogeneous, and resources will not be used equally. The provision of resources and accessibility alone cannot ensure improved mobility of vulnerable individuals (Martens et al. 2019; Pereira et al. 2017). In this respect, the transport disadvantages debate should explicitly consider any form of discrimination and marginalisation while acknowledging the needs and abilities of citizens who are vulnerable to exclusion (Kuttler and Moraglio 2020).

The transport disadvantages debate is increasingly interested in the process of digitalisation (Durand et al. 2021) because it is transforming current systems and enabling the emergence of new services (Macharis and Geurs 2019). Cities have been addressing the challenges and opportunities associated with digital transport services, such as shared transport. As Anderson et al. (2017) argue, shared mobility offers the opportunity to improve the mobility of vulnerable populations. However, to ensure that vulnerable populations benefit and use such solutions, their requirements, abilities, and motivations to travel must be thoroughly understood (Kuttler and Moraglio 2020). Moreover, new transport solutions should be tailored to the needs of users (Bierau-Delpont et al. 2019). Therefore, it is necessary to assess to what extent different social groups benefit from such services and if they are protected from the burdens that services may cause

(Martens et al. 2019). This assessment requires the identification of the groups that are more vulnerable to transport disadvantages and potential forms of exclusion.

Although it should be kept in mind that individuals may belong to several groups that are vulnerable to exclusion and therefore suffer from several forms of disadvantages (Jeekel 2018), the existing literature offers useful approaches to systematically distinguish such groups. Aspects such as age, gender, ethnicity, income, education levels and residential location have an impact on the disadvantages experienced by citizens when using digital transport services (Durand et al. 2021; Venkatesh et al 2012). Church et al. (2000) denoted seven elements of the transport system that contribute to the exclusion of certain populations: physical exclusion, which refers to physical barriers; geographical exclusion, concerning the residential location of users and the availability of services in that area; exclusion from facilities, highlighting the distance to key facilities; economic exclusion, concerning monetary cost; time-based exclusion, which refers to constraints related to working hours and schedules; fear-based exclusion, concerning fears for personal safety; and space exclusion, highlighting security or management of the space, which prevents access of certain groups. Currie and Delbosc (2016) listed six main forms of deprivation that might result in forms of disadvantages concerning shared transportation. These include the lack of information, money, support, security, adapted design, appropriate operating practices and self-confidence. Furthermore, Goodman-Deane et al. (2022) identified seven groups that are defined by some of the characteristics previously mentioned. However, they do not refer to ethnicity and highlight two additional characteristics that define vulnerable groups: having a migration background and a disability.

Age-related disadvantages are identified as being especially problematic for DSMS. This is because older citizens face several barriers when using digital solutions (Harvey et al. 2019; Pangbourne et al. 2020). Firstly, they are often more reluctant to try and adopt new technologies that they are less familiar with. Secondly, a relevant portion of this group cannot drive a car or no longer benefits from the same level of physical ableness as younger adults. This hampers the use of certain services or requires the adaptation of DSMS.

The aspect of gender proves to be relevant when identifying vulnerable groups. Several studies show how women benefit less from shared transport services and face more disadvantages than men, especially in developing countries (Durand et al. 2021; Zhang et al. 2020; Wiegmann et al., 2020). Similarly, ethnicity correlates with greater deprivation of transport services (Golub et al. 2019) which as van Egmond et al. (2020) argue is mostly related to income, discrimination and cultural preferences. Moreover, women, sexual minorities and certain ethnic minorities are, for instance, more likely to face additional forms of disadvantages as they might potentially suffer from harassment while travelling (Martens et al. 2019).

Income plays another important role because material deprivation is generally associated with low levels of engagement with digital technologies (Longley and Singleton 2009). Moreover, it has been identified that people with lower incomes, who often do not have a bank account and do not own a credit card, are less likely to own digital devices, have access to a reliable internet connection or be able to do online payments (Sherriff et al 2020). Likewise, the level of education is related to income, producing similar disadvantages in addition to the potential difficulties related to understanding information necessary to the use of DSMS. For instance, Wiegmann et al., 2020 found that the average car-sharing user in Brussels is highly educated.

The residential location might play a crucial role in the use of DSMS and the related benefits for citizens. The type of region and built environment will considerably limit the offer of such services. For instance, peri-urban or rural regions tend to host fewer transport options and, similarly, ICT infrastructure is less reliable and present in rural regions (Malik and Wahaj 2019). Moreover, residential location correlates with some burdens citizens face, such as air and noise pollution, or accidents (Martens et al. 2019).

As highlighted by Goodman-Deane et al. (2022), having a migration background might result in barriers related to language and cultural differences, and the transportation needs of people with a migrant background may also vary. The last characteristic that may result in a form of disadvantage and vulnerability is related to disabilities. Di Ciommo and Shiftan (2017) state that people with disabilities frequently experience difficulties and require assistance and additional information. Moreover, depending on the disability, physical access to the service and digital interfaces can be highly problematic (Reis and Freitas 2020).

4 A Framework to Thoroughly Understand Transport Disadvantages in DSMS

As explained in the previous section, increasing accessibility is not enough to overcome transport disadvantages, regardless of whether these efforts are aimed at obtaining equity or a minimum level for everyone. This is because transport disadvantages are related to complex social processes depending on factors not considered by egalitarian, sufficientarian or prioritarian approaches. Moreover, all three approaches might be oversimplifying, since the abilities and needs of people are heterogeneous and not everyone uses available resources in the same manner (Martens et al. 2019; Pereira et al. 2017). These approaches tend to be problematic in that they require assumptions about an acceptable level of inequality or a minimum level of accessibility (Kuttler and Moraglio 2020). Therefore, the study of shared transport from the perspective of transport disadvantages and social exclusion requires the use of a more comprehensive approach. A fourth normative approach, the Capabilities Approach (CA) (Nussbaum and Sen 1993; Sen 1979, 2005, 2009) could help to overcome the blind spots of the egalitarian, sufficientarian or prioritarian approaches.

The CA shifts the focus from 'resources' to 'capabilities', arguing that all individuals should enjoy a level of 'capabilities' which allow them to fulfil their needs and develop their lives (Luz and Portugal 2021; Pereira et al. 2017). For Nussbaum and Sen (1993), the focus on the distribution of resources overlooks the diversity of preferences and needs of individuals. Resources are not ends in themselves, but rather means to achieve aims. Therefore, the CA builds on the assumption that the most important dimension of life is the freedom of individuals to choose how to lead their life (Ryan et al. 2015).

The freedom of choice and agency considered by the CA are understood through five main concepts: resources, conversion factors, capabilities, choices and functionings

(Vechio and Martens 2021). Sen (1992, 2009) defines 'resources' as tangible and intangible goods and commodities available to a person, while 'conversion factors' are the social, cultural, environmental and personal context that frame and limit the possibilities of an individual. 'Capabilities' are sets of opportunities and freedoms available for people to choose and act, which are related to their resources and conversion factors. Sen (1992) defines 'choice' as the decision of a person in favour of a particular thing over another, and 'functionings' are what an individual actually achieves when putting their choices into practice and exercising their capabilities (Vechio and Martens 2021).

The concepts of 'conversion factor' and 'choice' help to understand that the capacity of each individual to use a resource for a specific objective will highly vary. The CA investigates the process of converting a 'resource' into a 'functioning'. It considers 'capabilities' a prerequisite to reaching opportunities and enjoying freedoms, which enable individuals to achieve their aims (Sen 2009). Furthermore, the CA assumes an adequate or minimum level of 'capabilities' exists that all individuals must enjoy. However, this assumption is challenged by the difficulty to establish such a minimum level and the fact that 'capabilities' are related to personal attributes such as gender, ethnicity, level of income, age and education (Kuttler and Moraglio 2020).

In the main theorisations of the CA, mobility is simply described as the ability to move from one place to another (Nussbaum 2000), without any explicit mention of transportation. The approach does not incorporate a thorough understanding of mobility, such as the one found in mobility studies (Urry 2007). However, the CA has been increasingly used in transport studies in recent years, having been incorporated by several researchers from different perspectives (Beyazit 2011; Flamm and Kaufmann 2006; Martens 2016; Pereira et al. 2017; Ryan et al. 2019). For instance, Banister (2018) argues that the CA is a relevant approach to studying transport inequality as it does not focus on maximising the potential mobility of people but rather on satisfying the choices and objectives of individuals. Likewise, many scholars state that the CA is the most adequate fairness approach to understanding the complexity of transport networks (Martens 2016; Pereira et al. 2017; Vecchio and Martens 2021). It takes into account various important elements: the diverse needs and motivations of individuals, how people interact with the transport system, and the resources at their disposal to reach opportunities depending on their characteristics and choices (Luz and Portugal 2021; Vecchio and Martens 2021). Furthermore, the adoption of the CA in transport studies offers the opportunity to move beyond traditional socio-technical perspectives and bring into the debate the cultural dimension of transportation.

An example of how the CA has been applied in transport studies is the work of Smith et al. (2012), who studied the transport disadvantages encountered by rural households compared to urban inhabitants. Likewise, Cao and Hickman (2019) used the CA to study the different uses that Beijing inhabitants make of metro line 1 depending on their socioeconomic characteristics and geographical location. Concerning shared transport, Sherriff et al. (2020) applied the CA to study the use of dockless shared bikes in Manchester and identified how personal and social conversion factors play a role in the use of such services. Hence, a range of diverging perspectives has emerged on how to apply the approach in practice, with two main strands of literature that diverge in what the concept of capability refers to.

The first strand of literature conceptualises the 'capabilities' as the ability of individuals to be mobile (Beyazit 2011; Flamm and Kaufmann 2006). From this perspective, a 'functioning' is the exercise of mobility, which is influenced by the context of individuals and limited by the skills and knowledge they possess. Kaufmann (2002) incorporates this perspective through the concept of 'motility', defined as the way in which individuals appropriate the range of possible actions concerning their mobility. The second strand focuses on the study of accessibility as a capability, envisioning capabilities as the possibility of an individual to engage in a variety of activities outside their home (Martens 2016). This conceptualisation comprises the idea of mobility as the ability to move through space. Herein, mobility is considered as a means to achieve an objective and not as an end in itself. From this perspective, a functioning is the exercised participation of a person in such activities, focusing on the person's ability to convert resources into participation in activities (Ryan et al. 2015; Vecchio and Martens 2021).

This second approach seems more adequate to convey the main theorisations of the CA, which revolves around the freedom of each person to develop their life. Moreover, since the concern of research on transport disadvantages and related social exclusion is not only to ensure people's mobility but rather that they participate in society and reach opportunities, this second perspective lends itself better to transport research from the point of view of social inclusion (Luz and Portugal 2021; Pereira et al. 2017).

If the aim is to enhance accessibility as a means to guarantee individual freedom, the focus should be to guarantee each individual an adequate level of access to essential activities that are necessary to meet basic needs and enjoy opportunities. Nevertheless, this does not entail that everybody benefits exactly from the same level of transport resources. Hence, traditional approaches that only focus on providing more resources to increase overall levels of accessibility might overlook the ability of individuals to convert resources into capabilities (Ryan et al. 2015). In this regard, the definition of accessibility, which in transport research is generally labelled as the physical access to goods, services and destinations, is repurposed by the CA. For instance, Pereira et al. (2017) consider accessibility as an individual attribute resulting from the interaction of personal characteristics, such as age, gender, socioeconomic conditions and ableness, with the person's environment, and sociocultural context. The literature that adopts this perspective is interested in how different social groups can participate in activities, studying the levels of accessibility of vulnerable groups, such as the elderly (Ryan et al. 2019), children (Borgato et al. 2020), ethnic minorities (van Egmond et al. 2020), low-income groups (Borgato et al. 2020; Cao and Hickman 2019), and people with impairments (Reis and Freitas 2020).

5   Adapting the Capabilities Approach to Appraise DSMS

As a result of the advent of digital transport services, studies on accessibility have increasingly incorporated the digital divide. Digital exclusion occurs when a person cannot appropriately use app-based transport solutions due to the lack of digital connection, the availability of a necessary device or the lack of digital skills (Groth 2019). Digital exclusion has become central to understanding the unequal use of digital transport services, such as DSMS, raising the concern about how digitally illiterate individuals could benefit from these solutions. As previously explained, vulnerable populations which already suffered transport disadvantages, often also face digital-divide exclusion, resulting in additional difficulties for vulnerable users and creating new forms of deprivation (Durand et al. 2021). Thus, Luz and Portugal (2021) incorporate the digital-divide exclusion into their definition of the CA.

As a continuation of the research mentioned in previous paragraphs, such as the work of Kaufmann (2002) and Luz and Portugal (2021), this paper proposes a framework to appraise DSMS through the lens of the CA. With this contribution, we aim at enabling a comprehensive understanding of the inclusivity of DSMS, a knowledge gap identified in the literature. The novelty of this framework is that it operationalizes the theoretical grounds of the CA to better understand the inclusivity of DSMS, and the use that vulnerable groups can make of such services. This framework implies that a person's use of DSMS relies on three main factors (see figure): 'material access', 'skills' and 'cognitive appropriation'. As shown in the figure, DSMS can be conveniently used when the three factors are met. Thus, when only two factors are met, the use of the services might be difficult or impossible. For instance, when an individual is lacking the necessary skills to use a service, the service cannot be instrumentalised, and when someone cannot cognitively appropriate the service, it will be unattractive to this person. Likewise, when there is no material access to a service, the service remains unavailable for users. Moreover, DSMS should consider these three factors to the extent to which such services will be useful for a person to freely fulfil an aim and reach a necessary destination (Fig. 1).

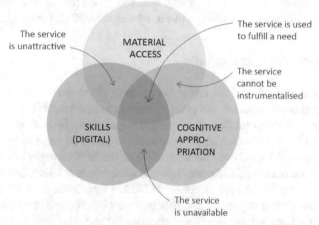

Fig. 1. Applying the CA to the study of DSMS.

The first factor, 'material access', refers to the 'resources' necessary to use DSMS, such as an available vehicle nearby, and the cost of use. Material access also refers to having a reliable internet connection and an adequate digital device, such as a smartphone or a tablet. In recent years, the smartphone has taken an increasingly central role in transport services (Gebresselassie and Sanchez 2018) with transport operators using a wide variety of applications that are often free. However, devices are not free of charge, and although there is available free wifi in some urban locations, having a reliable and permanent internet connection comes at a cost (Golub et al. 2019). Moreover, it is necessary to have an up-to-date operating system installed in the device, and enough battery or access to a charging point (Groth 2019).

The second factor, 'skills', refers to the 'conversion factors' that enable the instrumentalisation of a resource to fulfil an objective. This is because material access to technology does not ensure that someone benefits from a DSMS. Thus, 'skills', refer to the knowledge and the abilities necessary to use a DSMS, including the use of devices and applications. Vecchio and Tricarico (2018), argue that the skills necessary to use digital transport services are permanently evolving, and they can be differentiated into two types of skills: medium-related skills, which are related to operating a digital device, and content-related skills, which refer to information and strategic skills. The latter allows an individual to make strategic choices and select the most convenient information, route, services and use of their personal data (Durand et al. 2021).

The third factor, 'cognitive appropriation', refers to 'choices', which are informed by opinions, values, attitudes and motivations. Groth (2019) states that this factor is a crucial 'mental precondition' for individuals to engage with DSMS and identifies five dimensions that enable it: the autonomy experienced by users; the flexibility of the service; the excitement that the use of such service produces; the impact on social status perception; and privacy-related concerns. In this regard, Durand et al. (2021) define two main reasons that hamper the cognitive appropriation of an individual. The first one is related to a lack of trust in the technology, and a fear of security, reliability, and privacy, also highlighted by Harvey et al. (2019) and Groth (2019). The second one is due to the lack of desire or interest in the technology, either because the person does not know it or because the person does not want to use it, as stated by Zhang et al. (2020).

6   Discussion and Conclusions

Shared transport services are increasingly popular in cities around the world, allowing citizens to have short-term access to a vehicle, such as a shared car, bicycle or scooter, and potentially improving the mobility of vulnerable populations. Shared mobility providers mostly rely on digital technologies to operate their services, expecting users to learn and use their proposed app-based solutions. Thus, the lack of digital skills or internet connection and not having an adequate digital device, together with other factors related to digitalisation, may hamper the adoption of DSMS by a broader segment of the population. Not considering the needs and requirements of all social groups, may lead to transport disadvantages and deprivation, especially in the case of vulnerable populations. Nonetheless, the broader adoption of DSMS is not considered an objective per se, but a means to enable individuals to fulfil their needs more sustainably.

This work has identified existing approaches to the study of transport disadvantages to select an approach that can foster a better understanding of the needs and requirements of vulnerable populations concerning DSMS. Transport disadvantages are a complex social construct, and their study must consider the diverse characteristics of individuals. Therefore, aspects like gender, age, ethnicity, income, physical or cognitive impairments, education level and residential location must be taken into account by practitioners and researchers.

We consider the Capabilities Approach adequate because it goes beyond other approaches, not only looking at the availability of resources but also the capabilities of individuals. The CA argues that all individuals should benefit from a level of capabilities that allow them to freely fulfil their needs and develop their life, considering the needs of different social groups while acknowledging individual characteristics. This approach also diverges from the traditional perspective adopted to appraise transport services by going beyond socio-technical considerations and acknowledging cultural factors. Moreover, since the experience of individuals concerning DSMS is dependent on digital literacy, it is relevant that the CA incorporates the process of digitalisation.

Among other uses, this framework may be relevant to appraise the inclusivity of DSMS and facilities, evaluate uses among social groups, improve existing services, and orient policy-making. Likewise, this framework could also be used to appraise other transport services that comprise a digital dimension. By using predefined indicators, the three factors previously explained could be analysed. In order to facilitate the adoption of the framework by practitioners and policy-makers, a set of more concrete indicators related to study cases should be developed. Moreover, the framework needs to be integrated into existing working processes and should not require significant additional resources. From a research perspective, it is recommended to adopt qualitative methods, such as interviewing and focus groups, because the framework entails elements that concern complex socio-cultural phenomena.

Future studies could aim at identifying a standard set of indicators to operationalize this framework. For instance, analysing material access will require different data than studying skills or cognitive appropriation. The latter might be more difficult to grasp due to its intangibility and the fact that it is culturally embedded. Likewise, the lack of available data can be an obstacle to fully deploying the framework which considers personal characteristics and circumstances. Moreover, future research could seek to overcome the two main challenges of this framework. Firstly, the difficulty to fully incorporate the needs of vulnerable populations because such needs are the result of complex and multidimensional social processes. And secondly, to identify a possible minimum level of capabilities that should be facilitated to all individuals, by reducing the obstacles that impede their acquisition, and propose an adequate form of measurement.

References

Anderson, K., Blanchard, S.D., Cheah, D., Levitt, D.: Incorporating equity and resiliency in municipal transportation planning: case study of mobility hubs in Oakland, California. Transp. Res. Rec. **2653**(1), 65–74 (2017). https://doi.org/10.3141/2653-08

Banister, D.: Inequality in Transport. Alexandrine Press, Abingdon (2018)

Benenson, I., Martens, K., Rofé, Y., Kwartler, A.: Public transport versus private car GIS-based estimation of accessibility applied to the Tel Aviv metropolitan area. Ann. Reg. Sci. **47**(3), 499–515 (2011). https://doi.org/10.1007/s00168-010-0392-6

Beyazit, E.: Evaluating social justice in transport: Lessons to be learned from the capability approach. Transp. Rev. **31**(1), 117–134 (2011). https://doi.org/10.1080/01441647.2010.504900

Borgato, S., Maffii, S., Bosetti, S.: People on low income and unemployed persons. In: Re-thinking Mobility Poverty, pp. 124–134. Routledge (2020) te Boveldt, G., Keseru, I., Macharis, C.: Between fairness, welfare and feasibility: an approach for applying different distributive principles in transport evaluation. Eur. Transp. Res. Rev. **12**(1), 1–13 (2020). https://doi.org/10.1186/s12544-020-00428-4

Bierau-Delpont, F., Müller, B., Napoletano, L., Chalkia, E., Meyer, G.: Building an action plan for the holistic transformation of the European transport system. In: Müller, B., Meyer, G. (eds.) Towards User-Centric Transport in Europe. LNM, pp. 3–14. Springer, Cham (2019). https://doi.org/10.1007/978-3-319-99756-8_1

Cao, M., Hickman, R.: Understanding travel and differential capabilities and functionings in Beijing. Transp. Policy **83**, 46–56 (2019)

Casal, P.: Why sufficiency is not enough. Ethics **117**(2), 296–326 (2007). https://doi.org/10.1086/510692

Church, A., Frost, M., Sullivan, K.: Transport and social exclusion in London. Transp. Policy **7**(3), 195–205 (2000)

Cohen, B., Kietzmann, J.: Ride on! Mobility business models for the sharing economy. Organ. Environ. **27**(3), 279–296 (2014)

Currie, G., et al.: Investigating links between transport disadvantage, social exclusion and well-being in Melbourne – updated results. Res. Transp. Econ. **29**(1), 287–295 (2010). https://doi.org/10.1016/j.retrec.2010.07.036

Currie, G., Delbosc, A.: Transport disadvantage: a review. In: New Perspectives and Methods in Transport and Social Exclusion Research. Emerald insight (2016)

Davidsson, P., Hajinasab, B., Holmgren, J., Jevinger, A., Persson, J.A.: The fourth wave of digitalization and public transport: opportunities and challenges. Sustainability **8**, 1248 (2016). https://doi.org/10.3390/su8121248

Delbosc, A., Currie, G.: The spatial context of transport disadvantage, social exclusion and well-being. J. Transp. Geogr. **19**(6), 1130–1137 (2011)

Di Ciommo, F., Shiftan, Y.: Transport equity analysis. Transp. Rev. **37**(2), 139–151 (2017)

Dodson, J., Gleeson, B., Sipe, N.: Transport Disadvantage and Social Status: A Review of Literature and Methods. Griffith University, Australia, Urban Policy Program (2004)

Durand, A., Zijlstra, T., van Oort, N., Hoogendoorn-Lanser, S., Hoogendoorn, S.: Access denied? digital inequality in transport services. Transp. Rev. **42**(1), 32–57 (2021)

Feitelson, E.: Introducing environmental equity dimensions into the sustainable transport discourse: issues and pitfalls. Transp. Res. Part D: Transp. Environ. **7**(2), 99–118 (2002). https://doi.org/10.1016/S1361-9209(01)00013-X

Flamm, M., Kaufmann, V.: Operationalising the concept of motility: a qualitative study. Mobilities **1**(2), 167–189 (2006). https://doi.org/10.1080/17450100600726563

Gebresselassie, M., Sanchez, T.W.: "Smart" tools for socially sustainable transport: a review of mobility apps. Urban Sci. **2**(2), 45 (2018)

Gray, J., Rumpe, B.: Models for digitalization. Softw. Syst. Model. **14**(4), 1319–1320 (2015). https://doi.org/10.1007/s10270-015-0494-9

Golub, A., Satterfield, V., Serritella, M., Singh, J., Phillips, S.: Assessing the barriers to equity in smart mobility systems: a case study of Portland. Or. Case Stud. Transp. Policy **7**(4), 689–697 (2019). https://doi.org/10.1016/j.cstp.2019.10.002

Goodman-Deane, J., et al.: Toward inclusive digital mobility services: a population perspective. Interact. Comput. **33**(4), 426–441 (2021)

Groth, S.: Multimodal divide: reproduction of transport poverty in smart mobility trends. Transp. Res. Part A: Policy Pract. **125**, 56–71 (2019)

Harvey, J., Guo, W., Edwards, S.: Increasing mobility for older travellers through engagement with technology. Transp. Res. F: Traffic Psychol. Behav. **60**, 172–184 (2019)

Hodgson, F.C., Turner, J.: Participation not consumption: the need for new participatory practices to address transport and social exclusion. Transp. Policy **10**(4), 265–272 (2003). https://doi.org/10.1016/j.tranpol.2003.08.001

Jeekel, H.: Inclusive Transport: Fighting Involuntary Transport Disadvantages. Elsevier, Amsterdam (2018)

Jittrapirom, P., Caiati, V., Feneri, A.M., Ebrahimigharehbaghi, S., Alonso González, M.J., Narayan, J.: Mobility as a service: a critical review of definitions, assessments of schemes, and key challenges (2017)

Kaufmann, V.: Re-thinking Mobility: Contemporary Sociology, 1st edn. Ashgate, Aldershot (2002)

Kuttler, T., Moraglio, M.: Unequal mobilities, network capital and mobility justice. In: Re-thinking Mobility Poverty, pp. 39–48. Routledge (2020)

Lewis, E.O.C., MacKenzie, D., Kaminsky, J.: Exploring equity: How equity norms have been applied implicitly and explicitly in transportation research and practice. Transp. Res. Interdisc. Perspect. **9**, 100332 (2021)

Longley, P.A., Singleton, A.D.: Linking social deprivation and digital exclusion in England. Urban Stud. **46**(7), 1275–1298 (2009)

Lucas, K.: Transport and social exclusion: where are we now? Transp. Policy **20**, 105–113 (2012). https://doi.org/10.1016/j.tranpol.2012.01.013

Lucas, K.: A new evolution for transport-related social exclusion research? J. Transp. Geogr. **81**, 102529 (2019)

Lucas, K., Mattioli, G., Verlinghieri, E., Guzman, A.: Transport poverty and its adverse social consequences. In: Proceedings of the Institution of Civil Engineers-Transport, vol. 169, no. 6, pp. 353–365. Thomas Telford Ltd (2016).

Luz, G., Portugal, L.: Understanding transport-related social exclusion through the lens of capabilities approach. Transp. Rev. 1–23 (2021)

Machado, C.A.S., de Salles Hue, N.P.M., Berssaneti, F.T., Quintanilha, J.A.: An overview of shared mobility. Sustainability **10**(12), 4342 (2018)

Macharis, C., Geurs, K.: The future of European communication and transportation research: a research agenda. Region **6**(3), D1–D21 (2019)

Maffii, S., Bosetti, S.: Preface. Re-thinking Mobility Poverty: Understanding Users' Geographies, Backgrounds and Aptitudes (2020)

Malik, F., Wahaj, Z.: Sharing economy digital platforms and social inclusion/exclusion: a research study of uber and careem in Pakistan. In: Information and Communication Technologies for Development 2019: Strengthening Southern-Driven Cooperation as a Catalyst for ICT4D, pp. 248- 259 (2019). https://doi.org/10.1007/978-3-030-18400-1_20

Martens, K.: Transport justice. In: Transport Justice: Designing Fair Transportation Systems. Routledge (2016). https://doi.org/10.4324/9781315746852

Martens, K., Di Ciommo, F., Papanikolaou, A.: Incorporating equity into transport planning: Utility, priority and sufficiency approaches (2014)

Martens, K., Bastiaanssen, J., Lucas, K.: Measuring transport equity: key components, framings and metrics. In: Measuring Transport Equity, pp. 13–36. Elsevier (2019)

Meijers, E., Hoekstra, J., Leijten, M., Louw, E., Spaans, M.: Connecting the periphery: distributive effects of new infrastructure. J. Transp. Geogr. **22**, 187–198 (2012). https://doi.org/10.1016/j.jtrangeo.2012.01.005

Nussbaum, M.C., Sen, A.K.: The Quality of Life. Clarendon, Oxford (1993)

Nussbaum, M.C.: Women and Human Development: The Capabilities Approach. Cambridge University Press, Cambridge (2000)

Páez, A., Scott, D.M., Morency, C.: Measuring accessibility: positive and normative implementations of various accessibility indicators. J. Transp. Geogr. **25**, 141–153 (2012). https://doi.org/10.1016/j.jtrangeo.2012.03.016

Pangbourne, K., Mladenović, M.N., Stead, D., Milakis, D.: Questioning mobility as a service: Unanticipated implications for society and governance. Transp. Res. Part A: Policy Pract. (2020). https://doi.org/10.1016/j.tra.2019.09.033

Pereira, R.H., Schwanen, T., Banister, D.: Distributive justice and equity in transportation. Transp. Rev. **37**(2), 170–191 (2017)

Pereira, R.H., Karner, A.: Transportation equity. Int. Encycl. Transp. **1**, 271–277 (2021)

Reis, V., Freitas, A.: The predicaments of European disabled people. In: Re-thinking Mobility Poverty, pp. 147–161. Routledge (2020)

Ryan, J., Wretstrand, A., Schmidt, S.M.: Exploring public transport as an element of older persons' mobility: a capability approach perspective. J. Transp. Geogr. **48**, 105–114 (2015). https://doi.org/10.1016/j.jtrangeo.2015.08.016

Ryan, J., Wretstrand, A., Schmidt, S.M.: Disparities in mobility among older people: findings from a capability-based travel survey. Transp. Policy **79**, 177–192 (2019)

Santos, G.: Sustainability and shared mobility models. Sustainability **10**(9), 3194 (2018)

Sen, A.: Equality of what? Tanner Lect. Hum. Values **1**, 353–369 (1979)

Sen, A.K.: Inequality Reexamined. Clarendon, Oxford (1992)

Sen, A.: Human rights and capabilities. J. Hum. Dev. **6**(2), 151–166 (2005). https://doi.org/10.1080/14649880500120491

Sen, A.: The Idea of Justice. Belknap Press of Harvard Univ. Press, Cambridge, MA (2009)

Smith, N., Hirsch, D., Davis, A.: Accessibility and capability: the minimum transport needs and costs of rural households. J. Transp. Geogr. **21**, 93–101 (2012). https://doi.org/10.1016/j.jtrangeo.2012.01.004

Shaheen, S., Cohen, A., Zohdy, I.: Shared mobility: current practices and guiding principles (No. FHWA-HOP-16–022). United States. Federal Highway Administration (2016)

Shaheen, S., Cohen, A.: Overview of shared mobility. ITS Berkeley Policy Briefs, **2018**(01) (2018)

Sheller, M.: Theorising mobility justice. Tempo Soc. **30**, 17–34 (2018)

Schwanen, T., Lucas, K., Akyelken, N., Cisternas Solsona, D., Carrasco, J.-A., Neutens, T.: Rethinking the links between social exclusion and transport disadvantage through the lens of social capital. Transp. Res. Part A: Policy Pract. **74**, 123–135 (2015). https://doi.org/10.1016/j.tra.2015.02.012

Sherriff, G., Adams, M., Blazejewski, L., Davies, N., Kamerāde, D.: From mobike to no bike in greater manchester: using the capabilities approach to explore Europe's first wave of dockless bike share. J. Transp. Geogr. **86**, 102744 (2020)

Shibayama, T., Emberger, G.: New mobility services: Taxonomy, innovation and the role of ICTs. Transp. Policy (2020). https://doi.org/10.1016/j.tranpol.2020.05.024

Snellen, D., de Hollander, G.: ICT's change transport and mobility: Mind the policy gap! Transp. Res. Procedia **26**, 3–12 (2017). https://doi.org/10.1016/j.trpro.2017.07.003

Urry, J.: Mobilities. Polity Press, Cambridge (2007)

van Egmond, P., Wirtz, J.: Mobility poverty in Luxembourg. In: Kuttler, T., Moraglio, M. (eds.) Re-thinking Mobility Poverty: Understanding Users' Geographies, Backgrounds and Aptitudes, pp. 251–259. Routledge, London (2020). https://doi.org/10.4324/9780367333317-23

Vecchio, G., Martens, K.: Accessibility and the capabilities approach: a review of the literature and proposal for conceptual advancements. Transp. Rev. **41**, 833–854 (2021) https://doi.org/10.1080/01441647.2021.1931551

Vecchio, G., Tricarico, L.: "May the force move you": roles and actors of information sharing devices in urban mobility. Cities **88**, 261–268 (2018). https://doi.org/10.1016/j.cities.2018.11.007

Venkatesh, V., Thong, J., Xu, X.: Consumer acceptance and use of information technology: extending the unified theory of acceptance and use of technology. MIS Q. **36**(1), 157–178 (2012). https://doi.org/10.2307/41410412

Yigitcanlar, T., Mohamed, A., Kamruzzaman, M., Piracha, A.: Understanding transport-related social exclusion: a multidimensional approach. Urban Policy Res. **37**(1), 97–110 (2019)

Zhang, M., Zhao, P., Qiao, S.: Smartness-induced transport inequality: Privacy concern, lacking knowledge of smartphone use and unequal access to transport information. Transp. Policy **99**, 175–185 (2020)

Wiegmann, M., Keserü, I., Macharis, C.: Carsharing in the Brussels region. Brussels Studies. La revue scientifique pour les recherches sur Bruxelles/Het wetenschappelijk tijdschrift voor onderzoek over Brussel/The Journal of Research on Brussels **146** (2020)

Non-digital Travellers – Five Need-Based Personas to Understand Their Drivers and Needs

Suzanne Hiemstra-van Mastrigt[✉], Max Sampimon, and Claudia Spaargaren

Industrial Design Engineering, TU Delft, Landbergstraat 15, 2628 CE Delft, The Netherlands
S.Hiemstra-vanMastrigt@tudelft.nl

Abstract. Many people believe that low digital skills are only a problem of the elderly. However, the group of analogue or non-digital travellers is much larger and much more diverse than that. In the Netherlands alone, it is estimated that a group of 3–4 million people is not digitally able enough to make use of digital services. This is due to several reasons. In order to make use of digital mobility services, users need to be able and willing to use digital services. In transport, especially for demand responsive transport (DRT) services, the lack of digital skills can create a barrier for people to make use of the service. Based on insights from literature and interviews about digital skills, we have categorized the different groups of non-digital travellers, and created five need-based personas. On the basis of this, we formulated user requirements and design recommendations for mobility services, and for DRT services specifically.

1 Introduction

We live in an increasingly digital world, where an increasing number of services and routines are becoming "digital by default", including mobility services, like on-demand or demand responsive transport (DRT) services. If the rise of a digital infrastructure in public transport is accompanied by a disappearance of physical accessibility, the travellers' dependency on information technologies (IT) in transport increases. For travellers that do not want to or cannot cope with the digital transformations, this results in an increase of digital inequality, or even digital exclusion (Durand et al. 2019). This "digital barrier" can cause mobility poverty for this group, which might result in exclusion from the society (Durand et al. 2019; Sampimon 2020a).

1.1 Non-digital Travellers: A 'Forgotten and Unseen' Group

Many people associate low digital skills with elderly and because of this, they expect that this problem will solve itself in a few years' time. However, the group of analogue or non-digital travellers is much larger and much more diverse than that. Although the Netherlands ranks among the EU top in digital skills (CBS 2020a), 46% of individuals reported in 2019 that they have low (16%) or basic (30%) overall digital skills (Eurostat 2020). The average for individuals of the European Union (28 countries) with low or basic overall digital skills is even higher with 52% (26% and 26%, respectively). In the

I. Keseru and A. Randhahn (Eds.): *Towards User-Centric Transport in Europe 3*, LNMOB, pp. 74–92, 2023.
https://doi.org/10.1007/978-3-031-26155-8_5

Netherlands alone, it is estimated that the group lacking sufficient digital skills to make use of digital services consists of 3–4 million people.

The Digital Divide

The majority of people benefit from their digital skills. However, a rather large group of people benefits less and others not at all. This phenomenon of inequality is referred to as the digital divide (Van Dijk 2005). If public transport companies want to design inclusive services, the needs of these users have to be considered. However, there is not a clear view of who these non-digital travellers actually are; let alone, how to design for them. It seems that the transport industry has no clear sight of the travellers that might be affected by these digital transformations. The group that is often associated with low digital literacy is elderly. On the other hand, we see a hugely increasing number of elderly active on social media for instance (CBS 2020b), rising from 40% in 2014 up to 76% in 2019. According to a study by Durand et al. (2019), other factors are also associated with lower digital skills, such as a lower education degree, a lower income, being long-term unemployed, or low-literacy. Hence, since low digital literacy is not only associated with elderly, the digital divide will not be resolved over time. Therefore, it is of utmost importance to include and design for low digitally skilled in public services also in the future, including transport and mobility services.

1.2 Demand Responsive Transport (DRT) Services

Whereas fixed line transit (FLT) runs according to a fixed schedule and route, demand responsive transport (DRT) services operate on demand, and therefore, only when and where needed, making the service more efficient. DRT services typically involve users calling a booking service which plans a route for the day to pick-up users and take them to their required destination; best explained as a combination between bus and taxi (Mageean and Nelson 2003). Due to the proliferation of internet and GPS-enabled smartphones, operators are able to operate in real-time and on a large scale (Alonso-González et al. 2018). Additionally, the smartphone app allows the use of several more important features that significantly improve the user experience, such as getting notified about the existence of a DRT service, having the possibility to plan anywhere, getting a confirmation of the reservation, the possibility to locate the bus platform and to consult and being notified about (real-time) schedule updates, which minimises unnecessary waiting at bus station in case of changed departure times. This means that users who lack (the ability to use) these resources experience a – sometimes impregnable – threshold to make use of the service, leading to an increased risk of mobility poverty amongst the 'low digital skilled' (Sampimon 2020a). Hence, besides the benefits that the digitization of DRT services might bring to a majority of users, it also increases the digital divide by possibly excluding a large user group from using public transportation.

Likewise, Jittrapirom et al. (2019) presented the user perspective of a public shared on-demand transport service, where users can make a reservation using their smartphone or calling a helpdesk. Experts in their study considered this service not a viable option for the elderly as they can make no use of digital possibilities; *"to fully benefit from the available service functionalities, travellers are required to access the service via its app"* (Jittrapirom et al. 2019; p.6).

User Experience of DRT Services in the Netherlands

At the time of this study, there are 37 DRT services active in the Netherlands, operating under the umbrella of eight different operators. These services often get implemented in order to maintain public transport in low-demand areas. Many cases in the Netherlands show that occupancy rates have dropped once DRT has been implemented. It has been identified that there are several problems playing a role in these dropping rates (OV Loket 2019). In an earlier study by Sampimon (2020a), based on user interviews, online reviews and literature review, it was concluded that the majority of users find the implementation of DRT services a deterioration of the user experience compared to FLT services. Four of the main reasons for this are:

1. Travellers are not familiar with DRT services and therefore, they are not aware of the fact that a reservation is required.
2. It requires more actions by the traveller to initiate the journey and therefore, more effort.
3. The minimum reservation time in advance limits the spontaneity of the trip.
4. Departure and arrival times are often inconsistent. Insufficient feedback decreases the certainty of travelling.

These issues are often acceptable to travellers with access to the digital possibilities that are provided in smartphone apps from DRT operators. Journeys can be booked by means of the app, and real time travel information can be accessed via the same app. For non-digital travellers, however, who are lacking these possibilities, this makes their user experience of DRT services even worse.

For example, mobile phone users and smartphone users are able to perform the actions while travelling and let them plan and book their journey at the last moment, whereas the alternative physical resources, such as the landline and the desktop computer, are fixed in one place. Physical service desks are, except for one example, never incorporated in Dutch DRT services. This implies that the physical accessibility of the actions is limited, which drastically affects the spontaneity of the travel behaviour of non-digital travellers.

Consequently, users without a mobile- or smartphone are forced to plan ahead and are not updated on disruptions or detours. If travellers are used to FLT, this infringes spontaneous travel behaviour and perceived reliability.

1.3   Aim of This Study

The aim of this study is to contribute to defining and visualising this non-digital user group and getting an understanding of their needs and requirements, for digital services in general and specifically related to DRT services. The outcome can serve as an inspiration for the rapidly digitising mobility industry as a whole with the aim to help designing mobility services with the needs of non-digital travellers in mind.

2 Methods

First, the non-digital user is identified based on earlier studies, including the share in the population and their needs and requirements in the context of products and services. Secondly, the identified users were invited to take part in in-depth interviews to explore and discuss their travel and planning behaviour and digital skills.

The insights from the literature exploration, including the earlier benchmark of DRT services from Sampimon (2020a), and the user interviews, serve as input for the creation of the need-based personas and requirements for designing (DRT) transport services that are accessible to non-digital users.

2.1 The Digital Divide in Public Transport

The ladder model (Van Dijk 2005) explains how the digital divide applies to public transport and how it is recognized. Figure 1 shows the different levels of the extent to which an individual has access to digital possibilities and resources that lead to opportunities and tangible outcomes. Durand et al. (2019) adapted the model and added public transport related examples to it, such as journey planning. The third and highest level in the digital divide consists of users who actually enjoy tangible benefits from using the app, such as saving travel time. Everyone that is not on the highest step of the ladder has at least one thing in common, namely that they cannot access the tangible benefits from using the digital service. Some people who do not always have access and do not use it often or lack the skills needed to, are placed in the second level digital divide. Finally, an individual lacking the required materials (e.g., who does not have access to a smartphone) is located in the first level digital divide.

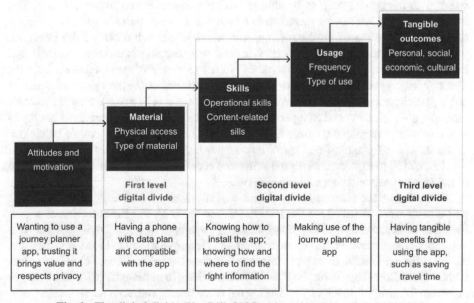

Fig. 1. The digital divide (Van Dijk 2005), adapted by Durand et al. (2019)

2.2 The Characteristics of Non-digital Travellers

The following characteristics are often related to low digital literacy (Durand et al. 2019; Stichting Lezen en Schrijven 2017a):

- Being elderly
- Having low literacy
- Having a migration background
- Having a low income
- Being long term unemployed
- Having a low education degree.

Low literates experience difficulties in developing digital skills because of language barriers, similarly to people with a migration background. People with a low income simply do not have access to material (and maintenance) because of financial reasons, but they also struggle to find the time to develop skills on a regular basis. People with low education degrees and long-term unemployed people are obstructed from developing skills and have limited motives to develop digital skills and therefore, are examples of lacking motivation (Van Dijk 2005). These insights give a better understanding of some factors that are causing digital barriers and the characteristics of the non-digital traveller. These characteristics have been used to recruit participants for in-depth interviews, to learn more about their needs and requirements.

2.3 Empathise with Users: Interviews About Travel Planning and Digital Skills

According to IDEO (2015), designers should immerse themselves in the context and speak with the target group to be able to design a desirable and fitting solution. The main goal of the interviews therefore is to empathise with the user groups. To get a deeper understanding of both conscious and unconscious behaviour of the interview participants, context mapping was applied. Context mapping is a design method that helps to inform designers about personal daily life experiences (Van Boeijen et al. 2020). Through, among other things, preparatory assignments before the interview, this method helps participants to be more aware of their experiences and to make intangible matters more tangible (Sanders and Stappers, 2008). The objectives of the interviews are to: 1) discover user perceptions towards DRT services; 2) reveal the user group's behaviour in the context of planning and travelling, which includes their preferred information sources, and 3) to explore how digital skills and access to the possibilities affect the use of public transport and how this is perceived.

Prompted by the characteristics of non-digital travellers described in Sect. 2.2, 12 participants were recruited for the interviews (see Table 1). For the recruitment of low literates, an organisation called 'Digi-Taalhuis' was approached. 'Digi-Taalhuis' is a volunteer organization that aims to help citizens improve their basic skills (language, digital skills, and math), in order to increase their level of self-reliance. Three volunteer ambassadors from Digi-Taalhuis (P10–12) were willing to participate (for a small fee) to represent the group of low literates. During the recruitment process of participants, a new group of non-digital travellers was discovered who are not yet described in earlier studies.

This group of -mainly young- users is characterized by their fundamental reluctance to use digital services (participants P7–9).

Table 1. Characteristics of interview participants

Participant	Gender	Age	Smartphone user	Target characteristic	Interview (via phone/physically)
P1	Female	50–65	Yes	Basic digital skills	Physically
P2	Male	50–65	Yes	Low digital skills	Physically
P3	Female	65+	Yes	Low digital skills	Via phone
P4	Female	65+	No	Low digital skills	Via phone
P5	Female	65+	Yes	Low digital skills	Via phone
P6	Female	50–65	Yes	Low digital skills	Physically
P7	Male	18–30	Yes, without data plan	Reluctant	Via phone
P8	Female	18–30	Yes, only banking and parking	Reluctant	Physically
P9	Male	18–30	Yes, only banking and parking	Reluctant	Physically
P10	Male	–	Yes, limited	Low literate	Physically
P11	Female	–	No	Low literate	Physically
P12	Male	–	Yes, limited	Low literate	Physically

For a context mapping session, it is preferred to observe participants and have a physical interview session. In times of the COVID-19 pandemic, observing the participants proved to be difficult due to social distancing measures and travel restrictions. Since the target group has low digital skills, a videoconferencing session was also not found to be a suitable alternative. Therefore, in case the interview could not take place at location, it was performed via telephone (landline) connection. This has resulted in slight differences in research setups of the interviews: sometimes at home via phone and sometimes at location (indicated in Table 1). Also, new insights arose throughout the interviews, leading to slight differences in the following interview sessions. Therefore, a semi-structured interview set-up was used. In this study, the outcomes of the context mapping interviews are used to develop personas.

2.4 Data Analysis: Creating Need-Based Personas

The insights from the different sources – such as desk research and user interviews – will be combined in the form of need-based personas, in order to provide a clear overview of the different types of non-digital travellers based on their needs instead of their demographic characteristics. The persona is a representation of a character including the shared needs and requirements incorporated in a medium, such as a descriptive story, anecdote, illustration or pictures. This medium allows the designer and other stakeholders to engage with the potential user groups and get inspiration for the design (Schneider and Stickdorn 2011). The goal is to make an engaging profile including the motives to barely or not at all use digital technologies and the specific needs of non-digital travellers (Schneider and Stickdorn 2011; Koos service design, n.d.).

First, a 2 × 2 matrix will be created that pairs two different sets of dimensions to distinguish several archetypical personas, or 'archetypes'. These dimensions are identified from the user interviews. Next, these archetypes are subdivided into the different need-based personas and described in more detail.

The outcomes (matrix and personas) will be presented and discussed in the following section.

3 Findings and Interpretation

In this section, the 2 × 2 matrix and need-based personas will be described, followed by design guidelines for DRT services and digital (mobility) services in general.

3.1 Can and Want Matrix: Four Archetypes

In order to use digital mobility services, users need to be 1) *able* and 2) *willing* to use digital services. In order to visualise this, a 2 × 2 "can & want matrix" was created (see Fig. 2). The Y-axis represents the ability to develop digital skills and understanding. The willingness to use (digital) technology is represented on the X-axis. Each quadrant represents an archetypical user group.

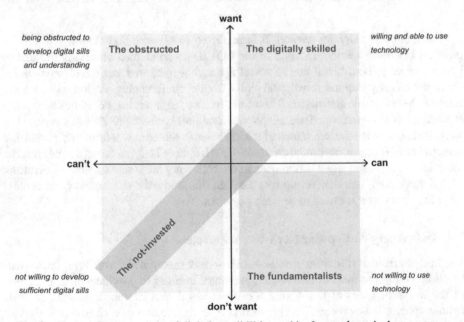

Fig. 2. The can & want matrix of digital possibilities and its four archetypical user groups

This leads to a differentiation of users. It consists of four core categories of users, called 'archetypes', which can be subdivided in different personas. Three of these archetypes are easily recognizable and visible: the obstructed, the digital skilled and the fundamentalist.

- The *obstructed* archetype (left in Fig. 2) consists of people that experience a barrier to developing digital skills nor understanding caused by a disability (e.g., low literacy or mental disorders, although the latter was out of scope for this study), which mostly derived from participants P10–12.
- The *digital skilled* archetype (top-right quadrant) consists of people that are willing and able to use digital technologies. They hardly experience any problems in accessing and using new digital services and are therefore out of scope of this study.
- The *fundamentalist* archetype (bottom-right quadrant) is defined by deliberately choosing not to use digital resources, which derived from participants P7–9.

However, the fourth archetype, derived from participants P1–6, is placed across the matrix and cannot clearly be fixed to one of the quadrants. This archetypical user group is defined by the people that have not invested in developing digital skills and therefore, they severely lack digital skills and understanding. The problem of this group of users, called "*the not-invested*", is that it might *seem* that they are digitally able; although in fact, they are not digitally able enough to keep up with the different developments.

3.2 Subdividing the Archetypes into Five Non-digital User Groups

Six different personas can be distinguished within the four archetypical user groups from the can & want matrix; five of which are non-digital users: the low literates, the conservatives, the low understanders and opportunists and finally, the digital detoxers (see Fig. 3). These five non-digital user types are briefly elaborated below and described in more detail and illustrated with a story in Sect. (3.3).

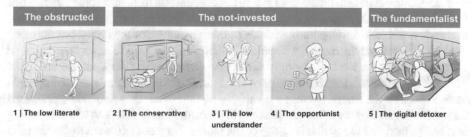

Fig. 3. Subdivision of the three archetypes into five need-based personas

The five types of non-digital users can be categorized in three archetypes: (A) the obstructed, (B) the not-invested, and (C) the fundamentalist:

A. The obstructed: Being obstructed to develop digital skills and understanding.

1. *The low literate*: Apart from whether they want to use it or not, low literates lack digital skills because they have difficulties reading.

B. The not-invested: Not willing to develop (sufficient) digital skills and therefore, severely lagging behind in digital skills and understanding.

2. *The conservative:* Not willing to develop digital skills.
3. *The low understander:* Often uses digital products and services, but has difficulties understanding them and therefore rather avoids them.
4. *The opportunist:* Can be very invested in their smartphone and using apps such as social media and messaging apps and their camera a lot. However, their knowledge is mainly situated in recreational apps, rather than functional apps, and might depend on help from family or friends.

C. The fundamentalist: Fundamentally against the use of a smartphone.

5. *The digital detoxer:* Deliberately choosing not to use a smartphone or data plan because they find it a degeneration of society and believe that they are better off without. Besides, they do not want to be obliged to use a smartphone and therefore, they are fundamentally against it too.

Referring back to the ladder model of Van Dijk, it seems that some people have chosen to never step onto the ladder of developing digital skills and are conservative in that regard. They cannot and they do not want to. Digital detoxers, however, stepped off the ladder at some point. Therefore, they do have the skills or are able to keep up with developments and maintain their skills, but they abstain from technology because of fundamental reasons. Basically, they can, but they won't use it. Then, there is a group that is simply not able to develop enough skills to take advantage of the digital possibilities; the low literates and the mentally disabled. Finally, this study has shown that someone using IT systems for work, is digitally skilled in performing the work-related tasks, but does not mean that this user is able to perform digital tasks in another field. These people have stepped on the ladder, but they are halfway on the ladder; they have a low or limited understanding or they are using it when the tangible outcome is clear and worth investing in. Finally, there is a group that is on top of the ladder. They can and want to develop and maintain their digital skills; the digital savvy. These are not part of the intended target group and therefore not taken into account in this study.

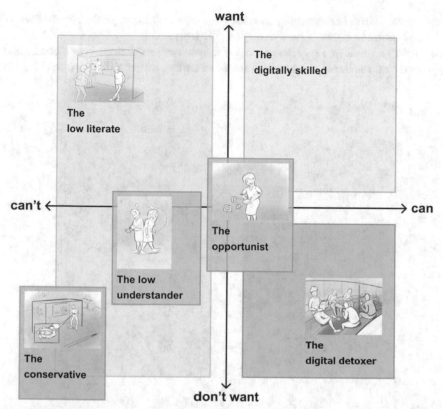

Fig. 4. The can & want matrix; an overview of the personas placed in the field of willingness over capabilities to use digital technologies

3.3 Five Need-Based Personas

In this section, each of the five non-digital personas from Fig. 4 is illustrated by a drawing and a story in the form of a quote. This quote-story includes the motives, needs and requirements of the specific character, thereby allowing the reader to empathise as if it were told by the actual user. These quotes are not taken literally from the interviews, but compiled (by the authors) based on the literature exploration and the interviews.

The Low Literate

"I often missed class during my time at elementary school, which caused a severe lack of literacy skills. I am ashamed of it and I'd rather not tell anyone about it. I am good at hiding it by avoiding any text related tasks.

I struggle with reading text above the language level A2 and writing myself. Especially, text becomes less accessible if it is long, small characters and many text boxes. When I have to fill in forms including writing my credentials and answering questions, I am likely to give up. This is already one disability that I have to cope with, but in the last decade another one arose: the digitisation. For me it is difficult to use a computer, tablet or smartphone, because I have low literacy skills and I have a low understanding of the digital world, like exploring and assessing information, digital safety and privacy. At home I have the internet, but I don't feel safe using it.

I prefer to speak to a real person, but if it involves text, it does help when texts are supported by visuals and photos of the actual situation, rather than just icons (Fig. 5)".

Fig. 5. Persona: The Low Literate

The Conservative

"I never felt the urge to understand any of the digital developments. Personally, I prefer working in the conservative way and this has worked for me. I don't see the benefits of using any digital option, so I won't use it. Back in the days, the world was much easier and nowadays, the world has become extremely complex. I am not ashamed of it and to be honest, I feel discriminated against if there is no analogue or physical option; you can't expect everyone to adapt by spending all their time developing digital skills. I don't even like it.

I prefer to go to an actual person at a counter and if not possible, I prefer to speak to them on the phone. I write down information on paper if I won't remember it or I ask for a brochure (Fig. 6)".

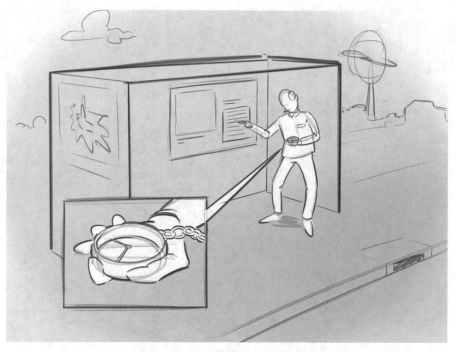

Fig. 6. Persona: The Conservative

The Low Understander

"I have very little understanding of computers. There was never really a reason to, but now I regret that I stepped into it too late. It has become too complex. It feels that it is the norm to use it and I feel silly that I just can't. When it comes to computers, I have a low self-esteem and I always think that I did it wrong. For some reason it does not stick and I don't understand any of the structure. I feel overwhelmed by text, many layers of information, log in credentials and other questions. This feeling is accompanied by being afraid to make irreversible mistakes. Especially when things turn red, then I know it is wrong and I want to start over. If I need to use a computer for some reason, I get anxiety. Besides, I wouldn't use computers, because I am unaware of the possibilities.

I prefer to fulfil the task without a computer involved. Preferably, I speak to an actual person, because then I have a real interaction. This allows me to ask questions that otherwise a computer is not able to answer. If I need to do a task online, I will get help from others. However, it feels as burden to them, so I prefer not to (Fig. 7)."

Fig. 7. Persona: The Low Understander

The Opportunist

"I use my smartphone for calling, using Whatsapp, to take pictures and several other basic stuff. My digital skills are limited to these tasks. I don't need more than that. I hate it if I need to download apps, unless someone else recommends it to me. If I have enough time and I am calm, I can learn any basic digital skill. However, I rarely need them so I only learn them if I clearly see the benefits outweighing the required effort. When an issue occurs that I am not able to cope with, I get irritated. I would try it myself first, but give up easily (Fig. 8)".

Fig. 8. Persona: The Opportunist

The Digital Detoxer

"I use a mobile phone to stay connected. Additionally, I perform quite a few tasks online. But – I deliberately choose not to use a smartphone in my daily life because it is a distraction and it makes me restless. I believe that it degenerates society because people get disconnected to each other and have less time for quality activities, such as reading a book or human interactions. Besides, it is unacceptable that people are required to use a smartphone for basic activities such as banking or transportation.

People can reach me on my mobile phone and if needed, I perform tasks online. Although, I appreciate it if there is a physical person available. After some time without using a smartphone, I start to lose my digital skills and knowledge (Fig. 9)."

Fig. 9. Persona: The Digital Detoxer

3.4   Goals for More Inclusive DRT Services

In order to provide a DRT service for non-digital travellers that is (more) acceptable and pleasant, the service should meet the following three conditions:

1) **Comprehensible information:** The non-digital user is aware of the existence and functionality of a DRT service before the journey and encouraged to make a reservation;
2) **Enhance certainty:** The non-digital user is certain about the departure and arrival times while travelling by providing real time route and time schedule information;
3) **Facilitate spontaneity**: The non-digital user is able to travel spontaneously.

3.5   Design Guidelines for DRT Services

Sampimon (2020b) describes a process for designing a Demand Responsive Transport service for non-digital travellers based on these three principles. Based on the DRT benchmark study (Sampimon 2020a) and user interviews, a set of guidelines can be drawn up that support achieving the previously described goals.

With regards to communication, it is recommended to explain the reason for implementing the new service. Transparency seems to be appreciated. Also, when introducing such service, the focus should be on the service itself, instead of focusing on the digital app. Communicating the beneficial features of the app on posters are perceived as noise and instead, they can be communicated within the app itself. In this way, non-digital users feel addressed, and information about digital features reach the people that actually use it. Also, when using a step-by-step explanation of the service, it is important that this is supported with real pictures instead of icons, supported by short readable sentences. Every piece of information should logically follow up on each other, with having clear visual dividers between them.

	The obstructed	The not-invested	The fundamentalist
Avoid too much text [level <A2]	!!		
Avoid the need to type	!!	+	
Avoid digital interfaces with layers of information, such as jumbles of questions and passwords	!!	!!	
Avoid long screen times	+	+	!!
Provide personal and verbal interactions	!!	++	+
Provide human interaction (via telephone or in real life)	!!	++	+
Be clear about the difference in online and offline possibilities	+	++	
Provide enough landmarks during the journey, such as directions, street names, departure and arrival times, and possible other relevant announcements	!!	++	
Use a non-computer like form style	++	++	
Provide instructions and feedback; Avoid red lights and colours	!!	!!	
Clarify the possibilities	+	++	
Payment by OV-chipcard	++	++	
The OV-chipcard may be incorporated in the design, but should not involve the requirement to include 'travel subscriptions'*	++	+	
Payment by bank card	++	+	
Provide an equal offline possibility next to the online possibility		++	+
Avoid the need to have a personal smartphone/data plan to use the service	+	++	!!

*A travel subscription or travel product is a certain feature that an individual can upload to their OV-chipcard in the Netherlands, such as discounts (e.g. for students or elderly), or access to specific services (e.g. use of OV-bike).

Fig. 10. Design guidelines for the different archetype users obstructed, not-invested and fundamentalist. Legend: light (+) = considerably improves the user experience for this group; medium (++) = significantly improves the user experience for this group; dark (!!) = without this, the service is unusable for this group.

Regarding the second goal (to enhance certainty), it is important to aim for providing offline alternatives to the app features. For example: provide real-time departure times at the bus stop, which includes a confirmation that the bus is actually coming to that specific bus stop. Secondly, do not only provide an app, but also a telephone number with personnel to make reservations. In the Netherlands, 0900-phone numbers (not toll-free) are intended for serious information and business services and were therefore preferred by participants because they are perceived as trustworthy. Also, because these numbers are easier to remember and recognizable.

3.6   Design Guidelines for Digital (Mobility) Services

Based on literature and outcomes of the interviews, some general design guidelines for digital (mobility) services were distilled and are summarized in Fig. 10. The impact on the user experience is indicated by the colour: the darker the colour, the more crucial this is to the user experience of this group (*note: this has not been validated but composed on the basis of the persona descriptions and interviews*). The classification in this table was made on the basis of the archetypical user groups: the obstructed (consisting of low-literates), the not-invested (consisting of conservatives, low-understanders and opportunists) and the fundamentalist (consisting of digital detoxers).

Even if smartphone apps are part of the service, considering these guidelines could increase the chance that low digital literates that own a smartphone will adopt the app. This concerns accessibility rules of copy and graphic design and interface design (in case digital features are part of the service), communication about the (new) service, payment, and ethics.

4   Discussion and Conclusions

4.1   Discussion and Limitations

The created personas are based upon earlier literature studies and 12 in-depth interviews. Although this is a relatively small number of participants, the diversity is large, so this gives insights into the needs and desires of this group. When performing qualitative research, such as context mapping, the number of participating users is usually small; 3 to 20 people (Van Boeijen et al. 2020), partly due to the fact that context mapping studies can be quite time consuming.

The majority of interviewed participants with low to basic digital skills are elderly females. It is assumed but not validated whether this is representative for this group although according to CBS (CBS Statline, 2013a), there is a preponderance of women among people over 65 years old (CBS Statline, 2013). The group of mentally obstructed or mentally disabled users was not part of the scope of this study and therefore not approached for an interview. Although they can be regarded as part of the Obstructed category, it can be questioned whether they would be able to travel with public transport independently.

Slight differences occurred to the interview set-up and questions due to the gathering of insights during the interviewing sessions and due to changing COVID restrictions and personal preferences in the middle of the COVID pandemic. Hence, some interviews took place at location but where this was not possible, participants were interviewed via the phone. However, the aim of the interviews was to gather rich, qualitative insights from the different types of non-digital users, to create personas that help to empathise and inspire, and that goal has been accomplished.

4.2 Conclusions

DRT services are often introduced as an efficient and thus more sustainable option to provide transport and to save costs. Nevertheless, public transport has its social function and therefore, it could be argued that investing in services that are accessible also for less digitally skilled people is socially responsible. Otherwise, this "digital barrier" can cause mobility poverty for this group, which might result in exclusion from the society (Durand et al. 2019; Sampimon 2020a).

Low digital skills are often associated with advanced age. This research has shown that the group of non-digital users is much larger and much more diverse. The use of need-based personas allows designers and other stakeholders to engage with the potential user groups and get inspiration for the design. The five need-based personas created in this study can help to empathise with the non-digital travellers when designing new (digital) mobility services.

Acknowledgements. This project received project funding for Public-Private Partnerships for Research and Development (PPP-allowance) from the Dutch Ministry of Economic Affairs and Climate Policy via CLICKNL, with contributions from the Dutch Ministry of Infrastructure and Water Management, TransLink Systems, DOVA, CROW, RET, GVB, 9292 Reisinformatiegroep and Rover. The authors would like to thank Dr. Euiyoung Kim and Prof. Hans Jeekel for their contributions to this project. Our thanks go out to all the participants, whose names will not be mentioned, for your time, effort and honest opinions and stories that you were willing to share.

References

Alonso-González, M., Liu, T., Cats, O., Van Oort, N., Hoogendoorn, S.: The potential of demand responsive transport as a complement to public transport: an assessment framework and an empirical evaluation. Transp. Res. Rec.: J. Transp. Res. Board **2672**(8) (2018). https://doi.org/10.1177/0361198118790842

CBS Statline: ICT gebruik van personen naar persoonskenmerken (2013). Accessed June 2020

CBS: The Netherlands ranks among the EU top in digital skills (2020a). https://www.cbs.nl/en-gb/news/2020/07/the-netherlands-ranks-among-the-eu-top-in-digital-skills. Accessed 14 Feb 2020

CBS: More elderly active on social media (2020b). https://www.cbs.nl/en-gb/news/2020/04/more-elderly-active-on-social-media. Accessed 20 Jan 2020

Durand, A., Zijlstra, T., van Oort, N.: Toegang geweigerd: digitale ongelijkheid in het slimme mobiliteitstijdperk. Paper Presented at Colloquium Vervoersplanologisch Speurwerk 2019, Leuven, Belgium, pp. 1–15 (2019)

Eurostat: Individuals' level of digital skills (until 2019). Data accessed via Eurostat Data Explorer (2020). https://appsso.eurostat.ec.europa.eu/nui/submitViewTableAction.do

IDEO: The field guide to human-centered design: Design kit (1st ed). IDEO (2015)

Jittrapirom, J., Van Neerven, W., Martens, K., Trampe, D., Meurs, H.: The Dutch elderly's preferences toward a smart demand-responsive transport service. Res. Transp. Bus. Manag. **30** (2019)

Koos service design: Need based personas (n.d.). https://www.koosservicedesign.com/tool/service-design-need-basedpersonas/

Mageean, J., Nelson, J.D.: The evaluation of demand responsive transport services in Europe. J. Transp. Geogr. **11**(4), 255–270 (2003). https://doi.org/10.1016/S0966-6923(03)00026-7

OV loket: Kwartaalrapportage 1 januari 2019 – 31 maart 2019 (2019). https://www.ovombudsman.nl/wp-content/uploads/KwartaalrapportageQ1_2019OVloketdef.-1.pdf

Sampimon, M.: The user experience of demand responsive (2020a)

Transport services. Research report, Delft University of Technology, Delft, the Netherlands, April 2020

Sampimon, M.: The design of a Demand Responsive Transport service for non-digital travellers. Design report, Delft University of Technology, October 2020, Delft, The Netherlands (2020b)

Sanders, E.B.-N., Stappers, P.J.: Co-creation and the new landscapes of design. CoDesign **4**(1), 5–18 (2008). https://doi.org/10.1080/15710880701875068

Schneider, J., Stickdorn, M.: This is Service Design Thinking: Basics, Tools, Cases. Wiley, Hoboken (2011)

Stichting Lezen en Schrijven: Factsheet laaggeletterdheid (2017). https://www.lezenenschrijven.nl/over-laaggeletterdheid/factsheets/factsheet-laaggeletterdheid-in-nederland

Van Boeijen, A.G.C., Daalhuizen, J.J., Zijlstra, J.J.M. (eds.): Delft Design Guide: Perspectives-Models-Approaches-Methods. BIS Publishers, Amsterdam (2020)

Van Dijk, J.A.G.M.: The Deepening Divide: Inequality in the Information Society. SAGE Publications (2005)

Methodological Paths to Achieve Inclusive Digital Mobility Solutions: Target-Group Capabilities and Limitations

Floridea Di Ciommo[1]([✉]), Gianni Rondinella[1], Yoram Shiftan[2,3], and Michelle Specktor[3]

[1] cambiaMO | Changing Mobility, Madrid, Spain
`floridea.diciommo@cambiamo.net`
[2] The Joseph Meyerhoff Chair in Urban and Regional Planning, Haifa, Israel
[3] The Israeli Center for Smart Transportation Research, Technion, Israel Institute of Technology, Haifa, Israel

Abstract. Physical, digital and graphic interface requirements of digital mobility and delivery services (and target groups) are a result of a comparison between the capabilities and limitations of each target group. A summary of the main users/non-users capabilities, limitations, and requirements (hereafter CLR) identified by populations that are more vulnerable will be the basis for understanding the most relevant needs threads: space, time and human factor. While space and time are traditional threads for capturing needs in transport and mobility (i.e. origin-destination, distance, time-saving etc.), the third thread 'human contact' appears as a new and clear need for the use of digital mobility and delivery solutions. A relevant number of inclusiveness requirements deals with this aspect that becomes a "must" for the extension of the inclusive digitalization in mobility. This chapter will conclude with the presentation of the most important insights in terms of capabilities, limitations and requirements that deal with the human contact factor.

1 Introduction

The aim of this chapter is to expose the use capabilities, limitations, and requirements (CLR) to the potential use of the Digital Mobility Services (DMS) and Digital Delivery Services (DDS) hereafter DMS/DDS that were found associated with each profile of the target groups of the INDIMO project. The concept of digital divide or digital exclusion was born, associated with the spread of digital tools for communication and organization of social life and the asymmetries of digital skills that actually exist among a variety of segments in the society. The digital divide is thus defined as the gap between those who have high access to digital tools and those who have low or no access at all, either because of not having access to the equipment, not having access to Internet connection, not having the adequate skills and capabilities or not feeling appealed by technology for doing everyday tasks in a different way (Saha 2014). A great part of the findings of the research hinges on the collective learning that was created during the

I. Keseru and A. Randhahn (Eds.): *Towards User-Centric Transport in Europe 3*, LNMOB, pp. 93–107, 2023.
https://doi.org/10.1007/978-3-031-26155-8_6

semi-structured interviews, that allowed us to get insights from various users and non-users belonging to the INDIMO target-groups such as they are specified in Table 1. The main findings explored throughout this paper show that for the five investigated pilots, digital technology, if it is not accompanied with human guidance and assistance, might be experienced as a barrier rather than as a facilitator in the use of the service. When digital applications do not address these adjustments for different target groups, traditional and learned paths to satisfy needs appear as the only alternative. It was found that certain populations have gained some familiarity with some specific apps (for instance, older people with WhatsApp, cognitive impaired youngsters with Instagram), but learned it in a very automatic and instrumental way. This does not mean that these persons are flexible in their approach to digitals tools in a way that allows them to explore new domains of digital knowledge by themselves. It was found, as seems clear in the cases of Madrid and Emilia Romagna, that the lack of familiarity with digital tools leads to different concerns associated with their use. These are mainly data privacy fears, fear of the lack of orientation or aid, the feeling of getting lost in the process or not being able to cope with so much information. When users are already familiar with digital services and when they are offered the adequate tools for guidance, including the possibility of contacting human assistance, the digital service opens up a wide range of alternatives, new behaviours regarding mobility and food consumption, new paths of autonomy and of self-confidence.

The next sections of this chapter present the methodology in Sect. 2, the insights from the fieldwork are presented in Sect. 3. Section 4 regroups the CLR paths in terms of time, space, and the human contact. The conclusion (Sect. 5) includes a reflection about the relevance of human contact in the era of digitalization, when the goal is to achieve inclusive digital mobility solutions.

2 Methodology

The data collection has been performed via in-depth semi structured interviews (SSI) that have been developed upon the INDIMO identified dimensions recalled in this section. The following figure provides a framework of data collection and analysis, through qualitative data gathered at each of the 5 pilots (Fig. 1).

The target-group respondents of users and non-users were the ones identified and included in the following user profiles corresponding to each pilot (Di Ciommo et al. 2022) (Fig. 2):

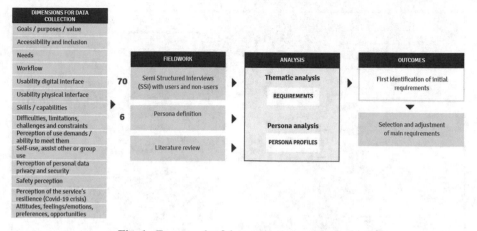

Fig. 1. Framework of data collection and analysis

Table 1. Pilots' names and user profiles

Pilot name and location	User profiles (and characteristics)
P1. Introducing digital lockers to enable e-commerce in rural areas (**Emilia Romagna**, Italy)	Older people and migrants/foreign people who receive/send parcels (lack of digital knowledge; residing in peri-urban or rural locations; lack of digital services; lack of dedicated network infrastructures; language barriers; low income)
P2. Inclusive traffic lights (**Antwerp**, Belgium)	Vulnerable pedestrian (older people; people with reduced mobility; people with reduced vision)
P3. Informal ride-sharing in ethnic towns (**Galilee,** Israel)	Informal ride-sharing users (ethnic minority; women; residing in villages or rural areas; language barrier)
P4. Cycle logistics platform for delivery healthy food (**Madrid**, Spain)	Delivery users (people with reduced mobility; people with reduced vision; people with mental health impairments; socially isolated-unwanted loneliness; not-connected people; low income; COVID-19 confined)
P5. On-demand ride-sharing integrated into multimodal route planning (**Berlin**, Germany)	On demand ride-sharing users (caregivers of children/ impaired/elders; women; lack of services; lack of digital skills, residing in peri-urban locations)

To enhance our knowledge and understanding focusing on users with physical impairments, the INDIMO partner MBE (Budapest Association of the Physically Impaired), conducted in Budapest, Hungary, a qualitative fieldwork of complementary interviews

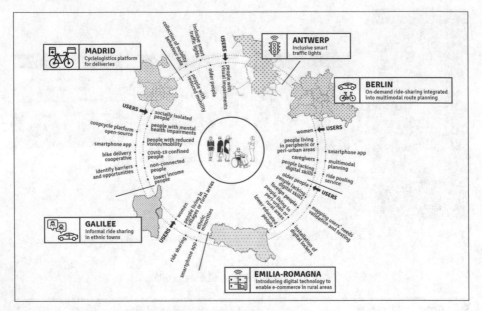

Fig. 2. The INDIMO pilots

to collect information about the public transport use of people with physical disabilities. This improved the focus on the specificities of impaired people with disabilities already included in the pilots in Antwerp and Madrid. Three user groups were selected: people with reduced mobility, people with reduced vision, and caregivers of people with disabilities.

Furthermore, to better understand the capabilities, limitations, and requirements of some of the addressed populations, it is sometimes needed to interview stakeholders, which are community organisations that work closely with the target population. Two different questionnaire templates were elaborated for both users and non-users interviews and a third one for stakeholder interviews.

For each interview, a debriefing document was filled in by interviewers based on a provided template. In the debriefing document the relevant fragments of each interview were included. In this way, the debriefing behaves as a summary with the highlights of the testimonies of the respondents. Afterwards, this text was used for the coding process and for moving forward with the thematic analysis (Rosala 2019).

The process from carrying out the interviews to coding and to identifying relevant themes included:

- the **coding process**: in which relevant verbatims from interviews are labelled with appropriate codes to identify and compare segments of text that are about the same thing. These codes allow us to sort information easily and to analyse data in terms of similarities, differences, and relationships among segments. The coding process has been conducted with the help of Quirkos CAQDAS (Computer Assisted Qualitative Data AnalysiS) software (https://www.quirkos.com/index.html)

- the **thematic analysis** is a systematic method of breaking down and organizing the identified codes for identifying and constructing significant themes (Rosala 2019).

3   Insights from the Fieldwork

The semi-structured interviews were focused on the problems and participants were eager to contribute and to find common solutions to common problems. The future approaches us at a high speed to face challenges regarding new social practices within the acceptance and usability of digital mobility services and digital delivery services.

We followed the CLR path (Capabilities, Limitations and Requirements), as the path that allows us to identify the requirements such as the difference between capabilities and limitations. These requirements are the inputs for the ulterior construction of the Universal Design Manual for digital mobility services. We organised the fieldwork and their instruments with the dimensions in accordance with the guidelines from INDIMO framework (Kedmi-Shahar et al. 2020), included in the below list:

Accessibility: Search for autonomy, reducing the dependency on relatives, friends or unknown passers-by in the street. Anticipation and control on the graphic interface are key elements to reduce the anxiety associated with orientation or excess of information in a digital environment. Real time input for users contributes to the feeling of continuous feedback and reassurance.

Inclusiveness: Human contact and assistance are a strong and constant element of this category of requirements. Human contact contributes to the warmth of relations as well as the feeling of flexibility and adaptation. The inclusion of different levels of digital competence, experience, language skills and socioeconomic status imply a strong need for adjustments and flexibility in the treatment. The availability of language options, but also a simple and familiar wording (using icons and images as part of this language) is also part of the requirement.

Additional Options: The DMS/DDS are seen not only in their current status, but also in their potentiality. Users pointed out the benefits that an extension of the delivered products, functionalities and services, including the covered geographical area, could bring.

Workflow: Most of these requirements address the simplicity with which the information is exposed, highlighted and treated. Requirements in this category deal with the ease of the navigation of the interface and the aids that this navigation may have for people who are not familiar with apps or who have specific difficulties. Be it the completing bar, the calculator and the error detection, these requirements target the feedback that the user has during the navigation process in order to ease the anxiety and reinforce the orientation.

Physical Interface: The interaction with the couriers or drivers generates a new layer of interface that is populated with its own reinforcements and barriers. The manners, help, offer and general friendliness of the service agents are highlighted. The way they express themselves, the introduction, the knowledge of the user's name and their identification contribute to the feeling of safety and trust building are key elements of these requirements. Also, new concerns arose about risk exposure in times of COVID-19 pandemics. In this context, an oriented training or the human contact availability can be useful.

Privacy and Data Security: This group of requirements is triggered by the sensitivity that some information (mainly bank and credit card information, address, phone number and personal identity) hold for the users. Transparency about the data that is stored and clearly conveying conditions of how the data can or will be used lay out the direction of these requirements.

Security and Safety: Especially sensitive for women, there are physical integrity concerns related to the interaction with rider/drivers, with the spatial setting and with the other users that may be part of the service. The requirements in this group tackle the effective response of the service to unforeseen situations of harassment, violence or assault specially related with gender.

Communications: Requirements in this category are related to the service exposing clearly their benefits and the target audience, expressing their social and environmental values in any, and facilitating the adoption and use through pieces of communication such as manuals, tutorials and lessons or the contact with facilitators.

COVID-19 Related: This category addresses the relevance in current times of working with clear protocols regarding the operation, which is especially relevant for people who feel more at risk in the face of pandemics, such as older people. These protocols should not only be in place but also actively communicated.

The main findings of the fieldwork could be grouped in the below categories (Giorgi et al. 2021):

Digital Gender Divide. A good part of the research on women and mobility focuses on the threats and the violence they face moving around in the public space. An important finding of our research shows that when women were parenting, their identities of mother stood above other identities, and their main concern was related to their children's safety, and about the interaction of others with their children. A main insight of our study shows that regular mobility services address a "male individual" user and do not contemplate the specific needs of caregivers in charge of dependents. This is an aspect of mobility that sometimes is obscured: transporting with others, either children or older people imply special requirements (type of vehicles, equipment, on-boarding and off-boarding spots etc.). Finally, women, especially when they are socially isolated, feel less comfortable with unproven technology.

Mobility and Physical Disabilities. A new insight of our study is that many people with reduced mobility are eager to show that they can have things done by themselves and may visualize the services of an app (for example, a service of food delivery) as an assistance that undermines their autonomy and their ability to solve issues on their own. Assistance appears as a two-fold aspect: as favouring autonomy or intruding in it; both as empowering and as a non-considered assistance. This has been long developed in our theme for Madrid, "Search for autonomy" and brings the focus on what levels of assistance are desired by different segments of the target-groups population.

Smart Traffic Lights and People with Impairments. The studies reviewed focused on smart traffic lights for the fluidity of vehicle circulation and there are not many articles that view smart traffic lights from the point of view of pedestrians. This way of thinking about the traffic is so rooted that, like it was found in the present research, vulnerable users incorporate this view when recognizing feelings of guilt for "stopping or delaying the traffic". This is a new insight that the present study casts light on. It was also found that there is no accessibility solution that is only a technological solution. In the case of Antwerp, if smart traffic lights were not accompanied by repairing and main-tenance works in the surroundings of the crossing, the innovation would be perceived as "just another gadget". This is a reminder to avoid the excessive techno-optimism and to bear in mind that digital approaches to problems always have a physical interface which has an important weight on the nature of the problem.

Foreign People as Central Public Users of DDS. The new insight of the present research is to identify the potentials of foreign people as central users of the locker system of parcel delivery. It was seen in the elaboration of the Emilia Romagna pilot that there is an unmet need of foreign people regarding the simplification of their exchanges with their families in their hometown. DDS offer a possibility of simplifying and enhancing this operation that is part of the life of someone settling down in a foreign country. Foreign people are presented in this way as potential users and participants of a new experience.

Non-connected People. One of the ways the present research goes beyond the bulk of the literature is that it does not consider all older people as a homogenous group. We found that many of the characteristics of low connectivity ascribed to older people were in fact idiosyncratic elements of specific contexts. This is the case of the examined rural areas where old mobile equipment (which blocks the possibility of a successful download of a new app) was associated with a more traditional mindset and the attachment to the "old way" of doing things. An idiosyncratic resistance typical of an environment goes far beyond the age cohort.

Most of the literature on user-centred approaches to include these groups mainly covers the feedback given by the app (through sounds, tones, pop-ups) but does not emphasize sufficiently the importance of the humans behind the digital interfaces, the need of direct contact with other humans, to give confidence and empower the user.

4　The Capability-Limitations-Requirements Paths for DMS/DDS

This section focuses on the identification of CLR paths. Given the specificities of each pilot and the fact that the services proposed are different in nature, this identification is provided pilot by pilot. Therefore, a summary of the CLR paths is presented. The organisation of the requirements per pilot and their target population profiles among various pilots and their points of contact concludes this section. Extensive work, both across time and space, allowed us to collect inputs from users, non-users and stakeholders of the target-group population associated with the design of digital mobility and digital delivery services (Di Ciommo et al. 2021a, b).

The Emilia Romagna Pilot 1 (P1) shows the differences of capabilities between both profiles of older people and migrants. The limited digitalization of older people makes them less confident with technology, while the low proficiency of migrants in speaking Italian and the discouraged use of tech for women by the patriarchal families represent a strong limitation for the adoption of this DDS, especially in the rural areas. While older people's requirements are oriented towards human assistance, migrants who have a strong need of the e-commerce service for exchanges with their home community, are asking for the availability of language choices in the digital lockers service. In both cases, the target populations have to overcome a cognitive limitation for using the services. Therefore, the solution will be focused on some specific training for both target populations. The below figure shows the CLR paths for pilot P1. The CLR paths, and concretely the requirements have been transformed in a clear recommendation of considering a human assistance for universally designing the digital mobility and delivery services in the future (Di Ciommo et al. 2021a, b) (Fig. 3).

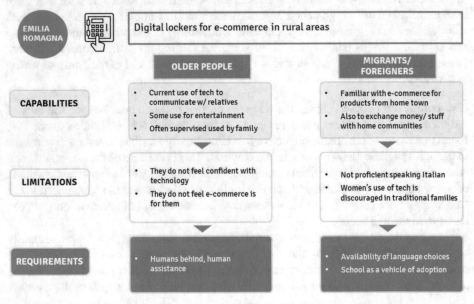

Fig. 3. Requirement path for the Emilia-Romagna pilot

The Antwerp Pilot 2 (P2) targets older people and persons with reduced vision and-mobility for the main limitations are related to the need for assistance when going to some unexplored place and the bad conditions of the public space, including road infrastructures. All three target populations require an extension of the duration of green light, while persons with reduced vision are asking as well for traffic lights with auditive signals and the communication of the status of lights (red/green). These two key requirements are at the basis of the recommendations for the INDIMO Inclusive Digital Mobility Toolbox including the Universal Design Manual for digital mobility and delivery services, the Universal interface language for digital transport services, the Cybersecurity and privacy assessment guidelines, and the Service and Policy Evaluation Tool. This pilot showed that the actions of policy makers in planning traffic lights and organizing public space infrastructure are equally relevant just like the digital app design development for satisfying the needs of end-users with some impairments. A consistent change of traffic and public space policy is required for shifting from a "car mandate" to a "care mandate" in mobility policies implementation. The below figure shows in detail the CLR requirements (Fig. 4).

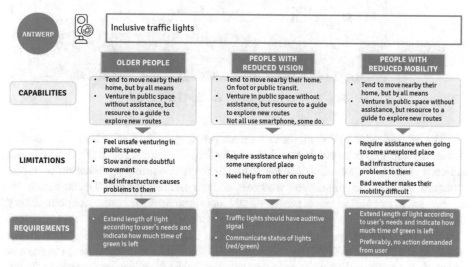

Fig. 4. Requirement path for the Antwerp pilot

The Galilee Pilot 3 (P3) shows that the DDS are already integrated in the life of Arab women in rural areas who already use the route planners and use the DDS for getting to work and school and gaining autonomy within a community with traditional ties. These capabilities are limited by the difficulty in reading a digital map, lack of coherence between the digital map and the real geography of the village, and the pressure of social mandates. Therefore, the women living in Arabic villages are asking for a stronger coherence between the digital map and languages, and the real-world geography and languages that will increase the community's confidence in the ride sharing digital mobility service. The below figure shows the CLR path for Pilot P3. The main recommendation for the universal design of the informal ride-sharing service deals with the community's confidence that can be increased if two key factors (i.e. ease to understand digital maps and coherence between digital maps and the real geography of the village) are considered. App algorithms should be based on an idiosyncratic development of the digital mobility service. Geography and space matters and should be considered in the digital development of the service (Fig. 5).

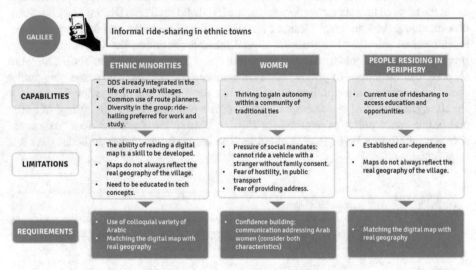

Fig. 5. Requirement path for the Galilee pilot

The Madrid pilot 4 (P4) shows that persons with impairments defend their level of autonomy and prefer to not focus on their physical conditions to justify mobility and delivery choices. However, the physical impairments constitute a limitation to access stores for people with impairment who need some human assistance. The current "non-connection of people" in respect to the digital tools increases the concerns about privacy and security that can be decreased through the simplifications of terms and conditions and the possibility of viewing user's ratings. The COVID confinement determined the need of establishing a COVID protocol for the DMS and DDS for avoiding the risks of exposure to the virus. The main recommendation for the inclusive and universal design of food delivery services is the simplification of the platform language in all its aspects from the terms and conditions to the words users need to understand to order the food. For example, common English words such as "courier" should be translated in the local language "repartidor". A total inclusive language approach should be adopted, as well as the possibility to reach human contact directly with the courier to arrange place and conditions of delivery (Fig. 6).

The Berlin pilot 5 (P5) shows the positive approach of women caregivers to the possibility of using this service when it has the right equipment and makes women feel comfortable and safe. Therefore, the main requirements include the possibility to have human contact and to arrange a place of pick-up or clear doubts. The geographical coverage of this service is a key requirement to be able to use it in a proper way. If digital ride-sharing services would shift from a male-oriented service to a universally accepted

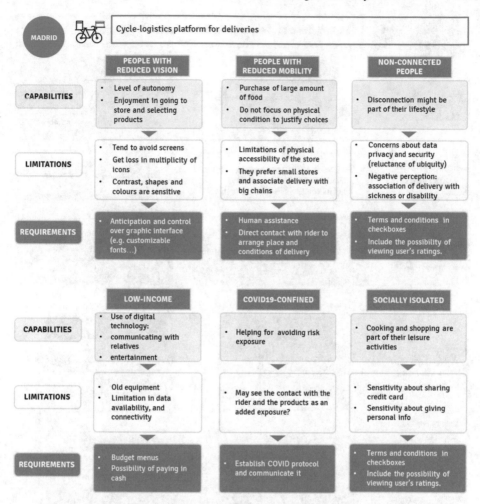

Fig. 6. Requirement path for the Madrid pilot

and adopted service by women, their "routine" and needs in terms of caregiving should be considered for universal design of the service. Time, space and human contact aspects should clearly be considered to move in this direction, as shown in the requirements included in Fig. 7.

The analysis of the capability, limitations and requirements paths show the nature of the needs for each pilot. Concretely, these needs can be grouped in three main categories: space, time and human contact, as highlighted in the table below and explained through the five different pilots. If space and time are two more classic dimensions of the mobility and delivery services, to pay attention to the human contact represents the novel factor to have a digital mobility and delivery solutions inclusive by design (Table 2).

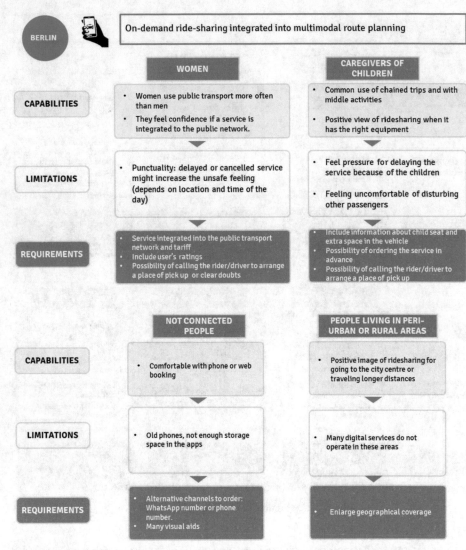

Fig. 7. Requirement path for the Berlin pilot

Table 2. Needs based on users' capabilities, limitations and requirements paths

Needs threads	Characteristics	P1 Emilia Romagna	P2 Antwerp	P3 Galilee	P4 Madrid	P5 Berlin
Space	Space is both a condition and a constraint to mobility The overlapping of spatial obstacles is a fundamental driver of mobility choices	Environment characterized by scattered rural villages Logistic problems linked to spatial configuration; involve a great amount of effort to pick up parcels: barrier to the satisfaction of needs	People with reduced mobility or vision find obstacles in the physical environment	Lack of adequate transport modes and connectivity in the Arab rural villages A hostile atmosphere prevents women to ride the public transit It is difficult to match the digital mapping with the real geography by Mobility apps	There are needs related to the geographical coverage of the service They affect people living in suburban areas who are most concerned with easy access to stores	People living in peripheral areas find problems with service coverage There are also concerns about the safety and attractiveness of the routes and the spots for onboarding
Time	Time is a valuable resource and the importance of making a good use of it appears in the different pilots	A locker for logistics allows a flexible and efficient use of time by the users	Extension of time to cross, the possibility of adapting time to target-group needs There's a different perceived time for each person	The app gives an orientation to time allocation: for instance, it makes universities and education centers closer to women	An app for food delivery may be time saving It gives a different quality to time: time to relax instead of time to cook; a gained time instead of a time devoted to a domestic task	Time needs to be flexible: (because children's needs are more unpredictable) And driver should be punctual

(*continued*)

Table 2. (*continued*)

Needs threads	Characteristics	P1 Emilia Romagna	P2 Antwerp	P3 Galilee	P4 Madrid	P5 Berlin
Human contact	Digital tools are something little familiar for a great variety of the groups Human contact is a requirement to overcome some of the fears contained in the digital domain	An assistant at the locker spot will be helpful to overcome digital-skills-related problems The importance of personal training is also remarked	People with reduced vision or mobility are depending on the help of passersby This assistance narrowed due to fears raised by the COVID pandemics	Having direct contact with the driver is a requirement to trust him, to overcome fears related with physical insecurity	The possibility of ordering food through WhatsApp or arranging details of delivery through a call to the rider were very frequents claims to the service	There was a request of humanity directed to the driver: women need drivers to care about the needs of a mother and to help her onboard and offboard

5 Conclusions and Recommendations

Some new insights of this paper that go beyond what was proposed by previous literature are related to the identification of specific needs of women. Most of the literature about women and mobility focuses on the gender-bias of transport planning and the negative experience of threat and potential harassment of women in the public space and transport. An important insight of the study is to show that regular mobility services address a "male individual" user and do not contemplate the specific needs of caregivers in charge of dependents, who are most of time are women. Concerning physical disabilities, a new insight is that assistance appears as a two-fold aspect: as favouring autonomy or intruding in it; both as empowering or as undesired assistance. Thinking of accessibility of street crossings, many papers have connection with fluidity of vehicles circulation and there are not many articles that view smart traffic lights from the point of view of the pedestrians, especially when they have physical disabilities. Finally, we also identify foreign people and migrants as central potential users of the locker systems of parcel delivery, for satisfying some of their unmet needs.

Based on these learnings, we have elaborated a list of inputs for the INDIMO Digital Mobility Toolbox, that may assist on the development and deployment of the digital mobility and delivery services of the future, and we have produced the main requirements for the digital and graphical interface of the apps, associated to the populations sensitive to them. In light of these requirements, we have developed a list of recommendations extensively included in the INDIMO deliverable D1.3 for developing the INDIMO toolbox and synthetized below.

Since the world has been transformed by the outbreak of COVID-19 and the exceptional situations that arose with it, the response and accommodation of different users' profiles to this anomalous situation was also explored. It was found that COVID-19 has a dual effect in most of the pilots: it may increase the need for apps to avoid a perceived mobility risk. But also, the new scenario may be experienced as a barrier to a new exposition which is contained in the use of service. The details of these findings will be examined in the remaining paragraphs.

1. Generally, users should be involved in the design and decision process before a new service is deployed. Developers, operators and policy-makers can better understand their target population with a participatory approach, such as in-depth semi-structured interviews. Integrate target populations, from diverse profiles, in the decision-making process about accessibility and inclusiveness of the digital services and apps. Only those who are genuinely concerned about accessibility can bring about changes in this area
2. To enhance the concept of human-centred design, it is advisable to start from the identified requirements in order to develop the design, technical, and visual solutions that address the aforementioned items. The CLR path (capabilities, limitations and requirements) allows design of- concrete profiles and with real users in mind.

References

Di Ciommo, F., Kilstein, A., Rondinella, G.: Intersectionality as a goal in data collection for mobility needs of social minorities. Transp. Res. Rec. J. Transp. Res. Board Res. Rec. TRR Paper-21-00604 (2022)

Di Ciommo, F., Rondinella, G., Kilstein, A.: INDIMO Deliverable D1.3 - users capabilities and requirements. INDIMO H2020 project (2021a). https://www.indimoproject.eu/

Di Ciommo, F., Kilstein, A., Rondinella, G.: D2.1 – Universal Design Manual – Version 1INDIMO H2020 project (2021b). https://www.indimoproject.eu/

Giorgi, S., et al.: Improving accessibility and inclusiveness of digital mobility solutions: a European approach. In: Black, N.L., Neumann, W.P., Noy, I. (eds.) IEA 2021. LNNS, vol. 220, pp. 263–270. Springer, Cham (2021). https://doi.org/10.1007/978-3-030-74605-6_33

Kedmi-Shahar, E., Delaere, H., Vanobberghen, W., Di Ciommo, F.: INDIMO deliverable d1.1 - analysis framework of user needs, capabilities, limitations & constraints of digital mobility services. INDIMO H2020 project (2020). https://www.indimoproject.eu/

Rosala, M.: How to analyze qualitative data from UX research: Thematic Analysis (2019). https://www.nytimes.com/2019/03/22/health/memory-forgetting-psychology.html. Accessed March 2020

Mobilising Local Knowledge:
Co-Creating User-Centric Services
and Applications

Leaving No One Behind: Involving Users in Creating Inclusive Digital Mobility

Kathryn Bulanowski[1]([✉]), Sandra Lima[1], and Evelien Marlier[2]

[1] European Passengers Federation, Ghent, Belgium
kathryn.bulanowski@epf.eu
[2] imec, Ghent, Belgium
evelien.marlier@imec.be

Abstract. With a shift towards the digitisation of mobility services, user involvement is vital for success. Especially critical is the inclusion of groups vulnerable to exclusion, so they can equally benefit from such services. In this respect, the Inclusive Digital Mobility Solutions (INDIMO) project established a multidisciplinary perspective on digital mobility services by considering the needs and concerns of vulnerable-to exclusion groups such as those who lack digital skills, belong to an ethnic minority or have reduced mobility.

Using data collection methods such as interviews and surveys targeted at vulnerable persons in five pilot locations, we collected information about user needs, intentions and preferences when using a digital mobility service. In this paper, we provide insights into the user recruitment process for this study and share tips for working with groups vulnerable to exclusion. Not to be forgotten are the lessons learnt from conducting this research during the COVID-19 pandemic.

1 Introduction

Digitalisation has produced many opportunities in the mobility sector, offering new services such as ride-sharing and journey planning, as well as services that provide more flexibility and access to real-time travel information. However, these digital services are not equally accessible to all members of society, especially to those part of groups vulnerable to exclusion such as people who lack digital skills, belong to an ethnic minority or have reduced mobility. User involvement is therefore vital to the success of any digital mobility service, so that these users are not left behind in an increasingly digitalised world.

The Inclusive Digital Mobility Solutions (INDIMO) project established a multidisciplinary perspective on digital mobility services by considering the needs and concerns of groups vulnerable to exclusion such as those who lack digital skills, belong to an ethnic minority or have reduced mobility. Through interviews and surveys targeted at vulnerable persons in five pilot locations, we intended to derive user needs, intentions, and preferences when using a digital mobility service.

Since groups vulnerable to exclusion are often hard to reach (Tovaas and Rupprecht Consult 2020), we focus in this paper on engagement with these groups who are often forgotten in the design of digital tools. We explore strategies for this purpose, and describe

© The Author(s) 2023
I. Keseru and A. Randhahn (Eds.): *Towards User-Centric Transport in Europe 3*, LNMOB, pp. 111–126, 2023.
https://doi.org/10.1007/978-3-031-26155-8_7

the methods used for recruiting participants and collecting data in the INDIMO interviews and surveys. Lastly, we highlight the results and lessons learnt of these engagement
activities.

2   An Increasing Digital Divide

Digital technologies have quickly become part of everyday life and have changed how
society functions on a daily basis. No exception to this digital evolution is the mobility
sector, which has been considerably impacted by technological advancements (Durand
et al. 2022, p. 33). There are many clear advantages to the digitalisation of mobility
services that allow for easier and more flexible travel, like real-time and location-based
information (Kuttler and Moraglio 2021). However, the risk exists that these digital
services generate inequalities because they may not be available or accessible to all
members of society. Digital services impose new requirements on potential users, and
people who lack resources, skills, autonomy, or a willingness to use new technologies
may be disadvantaged (Durand et al. 2022).

As digitalisation becomes more and more common and important in the modern
World, these inequalities become far more worrying for the future of our societies. The
digital divide can push groups into deeper levels of inequality and exclude them from
social participation. The term *"digital divide"* illustrates this inequality and concerns the
gaps formed between different societal groups in accessing and using information and
communication technology (ICT) (Saha 2016). Many factors can contribute to exclusion
from using digital mobility services like age, income, education, ethnicity, gender, and
even location (Durand et al. 2022). People with higher income, for example, often have
more access financially to digital tools as compared to people with lower income. They
often have access to the modes of payment required for these applications and they are
more likely to be able to afford these new services. Similarly, varying education levels
and cognitive impairments can also affect people's intellectual ability to utilise digital
technologies (Norris 2001). For example, digital mobility applications do not often take
into account neurodivergent user perspectives and even though, in the long-run, they
might create improvements to accommodate this group's needs, these add-ons might not
be as intuitive or user-friendly as needed for their actual use.

These factors do not act alone and often intersect to create even more vulnerable
situations. It is not uncommon to see older people who live in more remote areas, having
lower income and lower technological skills. In this scenario, we can see that they will
struggle not only from a lower offer of mobility (due to the remoteness of their homes)
but they will also have increased difficulties in using digital services that could possibly
increase their mobility options.

3   User Involvement: A Critical Step for Inclusivity

With the increasing shift towards digitalisation, (digital) mobility services must adapt to
the needs and requirements of all user groups, otherwise they risk further marginalisation
of already vulnerable groups. For this reason, their involvement in service design is
critical to ensure that these groups can access and use the services without facing any
barriers (Goodman-Deane et al. 2021).

One such process for ensuring that services benefit end-users is co-creation. This concept originates from the business and marketing sectors, as a method to collaborate on the design and production of products and services so that they better align with people's needs and wants. It has since reached global popularity for tackling challenges in many different sectors, through facilitating the active contribution of users (Puumala and Leino 2020). Though only recently applied to transport research, the co-creation process often combines different methods like interactive workshops and living labs, interviews, and even tests (Pappers et al. 2020). By involving users, these activities can help understand the barriers and challenges that they face in utilising a service or product.

Nevertheless, a first step not to be forgotten in this process is participant recruitment. Groups vulnerable to exclusion can be hard to reach for their involvement in such activities. "Hard to reach" groups tend to be underrepresented and are difficult to engage in public discussion (McCulloch 2020). This can stem from various characteristics including but not limited to:

- Demographic: such as place of residence, age or gender;
- Cultural: such as language or the lack of knowledge about how to become involved in these processes;
- Behavioural or attitudinal: such as the unwillingness to participate or distrust in government agencies;
- Structural: such as the lack of information in relevant languages or print sizes (Brackertz 2007).

Therefore, strategies for recruiting members of groups vulnerable to exclusion, that also take into account these characteristics, can help facilitate the co-creation process. Quite often, a tailored approach for each of the groups, that considers factors such as where to find them, which organisations they trust, and which networks they have contact with, can benefit the recruitment process and better engage potential participants (Brackertz 2007).

The INDIMO project funded by the Horizon 2020 programme of the European Union considers the needs, requirements and concerns of people who currently face barriers in accessing and using digital mobility services due to limited physical, cognitive or socio-economic factors. Subsequently, INDIMO utilised co-creation as a general tool for cultivating ideas and input from groups vulnerable to exclusion across five pilot locations[1]. The pilots and their different contexts are described below:

- **Pilot 1, Emilia-Romagna, Italy:** This pilot aimed to enable e-commerce (digital lockers) in rural areas and targets older people, foreigners and people with a low level of digital knowledge and education.
- **Pilot 2, Antwerp, Belgium:** This pilot created a Proof-of-Concept application that supports people with reduced mobility and visual impairments to safely cross intersections that are not equipped with accessible pedestrian signals.
- **Pilot 3, Galilee, Israel:** This pilot tested users' experiences and needs related to the ride-sharing mobile application, SAFARCON and focuses on people with limited

---

[1] For more information on the INDIMO project's analysis framework, consult the public deliverable Kedmi-Shahar et al. (2020)

access to mobility services due to their residence in the periphery or in areas with insufficient public transport. This pilot also targets people who lack digital skills or experience language barriers.

- **Pilot 4, Madrid, Spain:** This pilot tested users' experiences and needs related to the food delivery cycle logistics platform Coopcycle-La Pájara. It focuses on the needs of people with low-income, reduced mobility or vision impairments, as well as socially isolated and COVID-19 isolated persons.
- **Pilot 5, Berlin, Germany:** This pilot focused on an integrated ride-pooling service in Berlin. This pilot tests the experience of women and care giver users, while also considering the requirements of planning and booking multimodal journeys.

(Re)designing digital mobility services to take into account the specific needs of persons vulnerable to exclusion requires intensive field research, in addition to targeted strategies for recruiting participants. This paper therefore focuses on one aspect of the INDIMO co-creation process, a series of interviews and surveys which aimed at understanding the needs, intentions and preferences of people using a digital mobility service[2]. We report on the engagement behind these interviews and surveys, looking specifically at the strategies used for recruiting participants and collecting data, as well as the lessons learnt from these activities.

4   Involving Vulnerable Persons in the INDIMO Pilots

The activities described in this paper are part of a larger research project which has the goal of improving the understanding of users' needs in digital mobility services. The five pilots in the INDIMO project perform as an overarching platform for experimentation and applied co-creation as a general method. Through relying on the creative ideas and input of participants, we ensure that the project's results are based on real user needs, which will increase the rate of user acceptance in relation to digital mobility services.

The following sections focus on one part of this research; interviews and surveys in the five pilot locations that intended to understand how users would receive the introduction of new or redesigned digital mobility services in their communities. More specifically, we highlight the methods used for engaging participants for these purposes[3].

4.1   User Recruitment

As a starting point for selecting participants for the interviews and surveys, we referred to the INDIMO user personas created earlier in the project. Personas are a popular method for user-centred design, since they give more 'identity' to a user, as if they are real people (Harley 2015).

---

[2] For more information on this study, consult the public deliverable Marlier et al. (2021)

[3] All procedures performed in studies involving human participants were in accordance with the ethical standards of the institutional research committee of the Vrije Universiteit Brussel, the lead partner of the INDIMO project consortium (Ethics Commission in Humane Sciences of VUB, Reference number ECHW_238.02).

One user persona was created for each of the five INDIMO pilots, with the goal of representing the most relevant characteristics and profiles of potential end-users. The INDIMO personas were the results of co-creation activities within each pilot, like workshops with end-user representatives, end-users, policy makers and developers (Vanobberghen et al. 2021)[4]. A common characteristic of these personas is that they are all women, which acknowledges the fact that gender is an important factor to consider when designing inclusive digital mobility services. We especially put emphasis on the special requirements of women, because when new mobility services are developed and designed, they still often focus on men. An overview of the personas can be found in Fig. 1.

It is important to note that the INDIMO pilots recruited participants in the period of February – April 2021, during the second wave of the COVID-19 pandemic. For this reason, it was even more difficult to reach out to people belonging to the target groups for the surveys and interviews. These activities were normally going to be face-to-face activities, but because this was not allowed in that period, the researchers had to look for alternative methods. It forced some pilots to take more time to recruit citizens or to make the necessary online arrangements instead. For the Galilee pilot, the main challenge in finding participants was the limited use and exposure to the informal shared-ride app due to COVID-19. For the Emilia-Romagna pilot, it was difficult to reach the community of Monghidoro online. For the Antwerp pilot, response was limited because it was hard to find older people and persons with limited vision to participate in an online interview.

Furthermore, because most pilots found it challenging to reach the required number of 10 to 15 participants, we held a training session to support the recruitment of participants and to provide strategies for this purpose. During this session, they were provided with a number of concrete tips, including:

- Which channels the pilots can use to reach out to their target group;
- Which external parties/organisations can help with the recruitment;
- Which methodologies they can apply (e.g. snowball methodology)[5];
- Which incentives could be given to the participants;
- How they could shape the recruitment message (e.g. the importance to emphasise what is in it for them).

Table 1 displays an example of the strategies that were defined for the pilots.[6]

---

[4] For more information on the INDIMO persona creation process, consult the public deliverable: Vanobberghen, W., Vermeire, L., Giorgi, S., Capaccioli, A., Di Ciommo, F., Rondinella, G., Gabor Banfi, M., Tu, E., Lamoza, T., Spector, M. (2021). D1.2- User needs and requirements on a digital transport system. INDIMO project deliverable.

[5] In the snowball method, a participant or respondent is asked to identify other relevant persons to be involved in the research. See www.nsf.gov/bfa/dias/policy/hsfaqs.jsp#snow for more information.

[6] For more information on INDIMO's strategies for user involvement and co-creation activities, consult the public deliverable Royo (2020)

PERSONA
LUISA
Emilia Romagna
pilot

Age: 76 y/o
Marital status: Widowed
Children: One daughter
(+ one grandchild)
Occupation: Retired
Location: Centre of
Monghidoro
Income: Medium

PERSONA
JOHANNA
Antwerp
pilot

Age: 40 y/o
Marital status: Single
Children: No children
Occupation: Public
service officer
Location: Antwerp
Income: Medium

PERSONA
MARIAM
Galilee
pilot

Age: 25 y/o
Marital status: Not married
Children: No children
Occupation: Parttime
saleswoman at grocery store;
Parttime university student
Location: Rural area/village
Income: Medium

PERSONA
MARIA CARMEN
Madrid pilot

Age: 60 y/o
Marital status: Widowed
Children: Two children,
live on outskirts of city
Occupation: Unemployed;
support from government
and family
Location: Madrid
Income: Low

PERSONA
MARIE
Berlin
pilot

Age: 30 y/o
Marital status: Married
Children: Two children
(just gave birth)
Occupation: Maternity leave
Location: Peri-urban
location of Berlin
Income: Medium

Fig. 1. INDIMO user profiles

In the INDIMO project context, we targeted people who find it difficult to use digital services. Therefore, to further help with the recruitment, visuals and social cards were created that could be customised for each pilot location. By using this type of targeted communication, we aimed to address the people to whom the situation applies. Figure 2

Table 1. User recruitment strategies

Involvement strategy	How to implement it
Requesting support from institutional service agencies	Verify with the local administration which association could be involved to reach more people, preferably those institutions which have a relationship of trust with the target group. Share regular updates with administration and technical offices
Partnering with citizen groups, community and voluntary organisations	The number of associations in this area is limited. They could be contacted to verify their interest in involvement and then a suitable strategy could be applied (participate in their meetings, act as testimonials)
Activating snowball referrals	Select a first wave of users and ask them to spread the message to their peers
Announcing through media calls and advertisements	Verify what media our target users use and conduct a strategy accordingly. End-user representatives can help with pointing out the right dissemination channels
Engage through social media interaction	Verify if social media is a channel used by our target groups and study which interaction is most effective (campaign, ads...) to eventually implement it
Inviting a convenience sample through emails, newsletters, local papers and phone calls	Prepare a newsletter or email-template to be distributed. Municipalities can help with reaching out to their citizens, do not forget to involve them
Distributing flyers in places where users lie (shops, bars, pharmacies, etc.)	Leave leaflets in key points of the municipality, including the Post Office and City Hall
Ads on local classified websites (like subito in Italy)	This type of channel is more suitable for interactions among privates
Banner ads on most used apps by those groups	Verify which apps are most used by the groups, and implement an advertisement accordingly
Announcements on specific vulnerable groups magazines/newspapers/podcasts/radio programmes	The possibility to use these channels will be verified and a strategy devised accordingly

shows an example of a visual from the Antwerp pilot, which targeted older people, people with reduced mobility, and people with reduced vision. Figure 3 is an additional example from the Emilia-Romagna pilot.

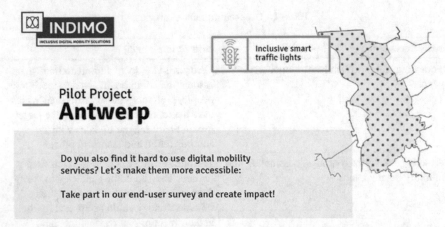

Fig. 2. Example of a visual created to help with the recruitment in Antwerp

Fig. 3. Example of a visual created to help with the recruitment in EMILIA-ROMAGNA

These materials could then be used on social media and on the partners' websites. Afterwards, each pilot team contacted the stakeholders or organisations needed for reaching potential survey participants in their local context. Some pilot teams also used their own communication channels to reach more people.

Furthermore, to incentivise people to take part in the study, we used:

- Gift vouchers
- Recyclable bottles
- Vouchers for drugstores/pharmacies

Specific strategies varied, based on factors like cultural inclinations or the limited penetration of digital communication methods. These are summarised in Table 2 below, along with the number of recruited participants, which were 90 in total.

Table 2. Applied user recruitment strategies

Pilot 1: Emilia-Romagna, Italy	
Pilot target group	Older people and migrants/foreign people, residing in peri-urban locations; lack of digital services; lack of dedicated network infrastructures; language barriers; low income, …)
Applied user recruitment strategy	The survey was published on two Facebook groups where many people living or interested in Monghidoro are updated on the local news; on the local partner's website; and was distributed by local stakeholders and organisations As an incentive for taking the survey, participants were given a gadget, which was sent in agreement with those who left their email contacts
Result	15 people participated. 40% were between the ages of 61 and 65 years old and 80% were women. 100% owned a smartphone and 86% owned a laptop. 66% could count on a strong social network and 66% had access to the bus or tram
Pilot 2: Antwerp, Belgium	
Pilot target group	Vulnerable pedestrians (i.e. older people; people with reduced mobility; people with reduced vision)
Applied user recruitment strategy	Stakeholder organisations that regularly work with the target audience of the survey were contacted and asked to help with the dissemination To incentivise the target audience, three participants could win a voucher for an online store if they completed the survey and registered for the lottery
Result	44 people participated; 80% were above 60 years old. 90% of the participants owned a smartphone and 80% owned a laptop. 79% had access to the car as a driver and 39% had severely limited support from care keepers
Pilot 3: Galilee, Israel	
Pilot target group	Informal ride-sharing users (ethnic minority men/women; residing in the periphery; language barrier; lack of digital skills)
Applied user recruitment strategy	A local feminist organisation directly reached out to participants via phone calls Participants volunteered to take part in the survey and did not receive any incentive for doing so
Result	5 people participated; all women ranging from 18–65 years old. All obtained a university degree or higher, owned a smartphone, and had access to the bus or tram

(*continued*)

Table 2. (*continued*)

Pilot 4: Madrid, Spain	
Pilot target group	Delivery users (people with reduced mobility; people with reduced vision; socially isolated-unwanted loneliness; not-connected people; low income; COVID-19 confined)
Applied user recruitment strategy	The questionnaire was distributed among participants in a locally-organised co-creation meeting Participants did not receive any incentives for taking the survey
Result	10 people participated. Most were in the 45 to 54 and 55 to 64 age groups. All participants had access to a smartphone as well as access to either a desktop PC, a laptop or a tablet. Most had access to a car as passenger
Pilot 5: Berlin, Germany	
Pilot target group	On-demand ride-sharing users (caregivers of children/impaired/elders; women; lack of services; lack of digital skills, residing in peri-urban locations)
Applied user recruitment strategy	A local partner offered support in contacting women with low incomes. As this is sensitive info, they could not pass the contact details to the INDIMO-partners. That is why they reached out to them themselves We offered participants a 30-euro gift voucher for their participation
Result	16 participated; all of which were women ranging from 18 to 60 years old. Many participants indicated having a good social network and most owned smartphones and laptops. 93% had access to the train or metro

4.2 Data Collection

After various meetings with the pilots and research partners on how to contact and engage the identified target groups, it became clear that there was no one-fits-all solution that we could apply to collect the necessary information. That is why various research methods were also used for the data gathering, with each method adjusted to the target group and the context of the specific pilot. An overview of the research methods used in the different pilots can be found in Table 3 below:

Table 3. Applied data collection strategies

Pilot target group	Applied research method for data collection
P1. Older people and migrants/foreign people, residing in peri-urban locations; lack of digital services; lack of dedicated network infrastructures; language barriers; low income, …)	Customised online survey in Italian Tool used: Survey Monkey Two versions were created: - One for people having used digital locker systems before - One for people that had never used such a service
P2. Vulnerable pedestrians (i.e. older people; people with reduced mobility; people with reduced vision)	Customised online survey in Dutch Tool used: Survey Monkey Two different versions of the survey: - One for people already using digital applications that assist them when they are travelling - One for people not making use of digital applications when they are moving around - Surveys had voice over function, so people with a visual impairment were also able to fill it in
P3. Informal ride-sharing users (ethnic minority men/women; residing in the periphery; language barrier; lack of digital skills)	Face-to-face interviews in Arab conducted by a local feminist organisation that focuses on empowering women Two different versions of questions were developed: - One for participants who have already used ride-sharing services in the past - One for participants that have never used a ride-sharing service before
P4. Delivery users (people with reduced mobility; people with reduced vision; socially isolated-unwanted loneliness; not-connected people; low income; COVID-19 confined)	Customised online survey in Spanish Tool used: Google forms One version was developed for people that are currently using a food/grocery ordering service

(continued)

Table 3. (*continued*)

Pilot target group	Applied research method for data collection
P5. On-demand ride-sharing users (caregivers of children/impaired/elders; women; lack of services; lack of digital skills, residing in peri-urban locations)	Use of paper format, so questionnaires could be filled in on-the-spot Local partner in Marzahn who offers support for low-income women, conducted the interviews. Later, the responses were integrated in the online tool 'Typeform' to enable the analysis Two versions were developed: - One for participants that already used ride-sharing services before - One for participants that have not used ride-sharing services before

We analysed the outcomes per pilot and formed connections to make overarching conclusions related to user engagement and data collection. The following section presents an overview of our results, including the lessons learnt.

5   Discussion

Collecting responses for this study was not always easy due to the COVID-19 pandemic, the holiday period, and limited digital exposure by some members of the targeted groups. For these reasons, it was challenging to get in touch with vulnerable groups and reach the target of 10 to 15 participants in each pilot. However, we still managed to recruit 90 participants with a diversity in age, education, and digital skill level. We collected data related to the following aspects (among others):

- Socio-economic status;
- Access to different mobility modes;
- Support from their social networks.

In the following sections, we highlight some lessons learnt from these activities, which can provide insights and helpful tips for engaging groups vulnerable to exclusion in similar data collection activities.

5.1   User Recruitment

In general, groups vulnerable to exclusion are often hard to reach (Tovaas and Rupprecht Consult 2020). The COVID-19 pandemic made it even more difficult to engage with and recruit groups vulnerable to exclusion in the INDIMO study, especially older people and people with a migration background. Some pilots required more time to recruit participants or to make online arrangements for an interview. For the Galilee pilot, for example, the main challenge in finding participants was the limited use of and exposure to the informal ridesharing app due to COVID-19. For the Emilia-Romagna pilot, it was difficult to reach the community of Monghidoro online. For the Antwerp pilot, response was limited at first due to the Easter holiday.

As a result, some pilots had a **bias in participants**. While this can be partly explained by the COVID-19 pandemic, it can also be explained by the fact that only one recruitment channel was used in some cases. By only utilising online channels for example, there is a good chance that only people with digital skills will be reached. Similarly, recruitment conducted via one main organisation, will also engage a homogeneous group of people. In the case of older target groups, participant recruitment via email generally reaches people with (at least some) digital skills, meaning that results are not always representative of the entire elderly society. That is why it is always important to **conduct recruitment through a variety of channels**, both on- and offline and to collaborate with different organisations. In the Antwerp pilot for example, participants were recruited through the channels of user representative organisations. These organisations have a network of people providing care to the target audience or have members that belong to the target audience because they provide services to support them. Because the recruitment message came from an organisation they trust, people were more willing to participate. In a similar light, the snowball referrals method for recruiting users can also be helpful to diversify participants, since the first wave of engaged users can spread the message to their peers.

Similarly, providing incentives, such as vouchers, discounts or small gifts, can also facilitate the recruitment process and increase the level of interest in participation. In this way, groups less likely to participate will be more inclined to do so, such as people with low income (Berlin et al. 1992).

Attractive images, visuals and storytelling techniques are also useful instruments for participant recruitment. Storytelling, for example, is a powerful tool that should be applied more often. As a popular trend in communication activities, storytelling can create emotional connections with an audience, which can increase engagement and even motivation (Love 2008). It can similarly attract attention to a research activity and provide context in a way that is relatable to the target group (Mathews and Wacker 2008). Looking for and using an element that is recognisable to the target group or audience helps them identify themselves in the situation being conveyed. In the INDIMO project, we used images of our targeted groups, like older people, in our social cards to attract and engage with them.

Furthermore, if circumstances allow, it also pays off to organise face-to-face events. By doing so, chances of obtaining more participant diversity are higher, and people feel more involved and in general are more willing to engage on a longer term.

To further empower recruitment teams, it can also be beneficial to **organise a recruitment training session** for sharing tips and tricks as well as **strategies on different user involvement techniques.** Recruitment teams will then have access to multiple methods that they can tailor to their needs and desired outcomes.

In the INDIMO context, we also found that participants and local organisations that helped with recruitment and data collection were also motivated to contribute to other upcoming research activities if we shared outcomes with them. In the Antwerp pilot, we organised a brief meeting to **share the most important findings** for the people that participated. This meeting was warmly welcomed. Some of the other external parties involved in the user engagement also indicated that the co-creation process was a **valuable learning process** for them as well. This highlights the importance of starting to question at the beginning of participation what the participants themselves wish to get out of the co-creation process. If they can benefit themselves, they will be all the more motivated to be more engaged.

5.2 Data Collection

Before disseminating a questionnaire, especially to vulnerable groups, it can be beneficial to ask at least one person of the target group to **proof-read the survey or questionnaire**. This will help detect if questions are difficult to understand or can be misinterpreted by the respondents.

Similarly, because the INDIMO pilots have their own contexts, target groups and cultural-specific aspects, they required customised approaches. This individual support was more time-consuming than expected, because additional time was needed to develop and create customised versions of the surveys, online tools and interviews, since they were tailored to each specific pilot context. It is therefore recommended to **foresee enough time for coordination and for the data gathering activities.**

6   Conclusion

Our ambition for involving vulnerable persons in the INDIMO co-creation process was to understand the barriers and drivers related to the use of inclusive digital mobility services. More specifically, we interviewed and surveyed 90 participants across five pilot locations to gain insights into their needs and requirements regarding digital mobility services.

Despite the strong efforts of the INDIMO pilots to engage with groups vulnerable to exclusion, we did not always manage to achieve the targeted number of 10 to 15 participants in each location. Furthermore, while we aimed to include a wide range of user profiles in these INDIMO surveys and interviews, we noticed that there was a bias in participants, and that the targeted groups were not always reached. This can be partly explained by the COVID-19 pandemic, and by the fact that sometimes recruitment was too one-sided through one specific channel. For example, only utilising online channels for participant recruitment can result in only reaching people with digital skills. It is, therefore, important to consider organising dedicated training sessions to share tips and tricks on different user involvement strategies, and to utilise a variety of channels, both on- and offline for recruitment and data collection. By doing so, chances of getting a more diverse set of participants and data are higher.

References

Berlin, M., et al.: An experiment in monetary incentives. In: Proceedings of the American Statistical Association (1992)
Brackertz, N.: Who is hard to reach and why?. ISR Working Paper (2007)

Durand, A., Zijlstra, T., van Oort, N., Hoogendoorn-Lanser, S., Hoogendoorn, S.: Access denied? Digital inequality in transport services. Transp. Rev. **42**(1), 32–57 (2022)

Harley, A.: Personas make users memorable for product team members. Nielsen Norman Group (2015). https://www.nngroup.com/articles/persona/

Frequently Asked Questions and Vignettes. National Science Foundation. https://www.nsf.gov/bfa/dias/policy/hsfaqs.jsp#snow

Goodman-Deane, J., Kluge, J., Roca Bosch, E., Nesterova, N., Bradley, M., Clarkson, P.: Digital mobility services: a population perspective (2021)

Kedmi-Shahar, E., Delaere, H., Vanobberghen, W., Di Ciommo, F.: D1.1 - Analysis Framework of User Needs, Capabilities, Limitations & Constraints of Digital Mobility Services. INDIMO project deliverable (2020)

Kuttler, T., Moraglio, M.: Learning Mobility. Re-Thinking Mobility Poverty: Understanding Users' Geographies, Backgrounds and Aptitudes, 1st edn., pp. 23–38. Routledge, Oxfordshire (2021)

Love, H.: Unraveling the technique of storytelling: taking advantage of a simple method to make your message stick. Strateg. Commun. Manag. **12**, 24–27 (2008)

Marlier, E., Lima, S., Bulanowski, K., Royo, B., Di Ciommo, F.: D3.4 - Report on user needs and requirements assessments in pilots. INDIMO project deliverable (2021)

Mathews, R., Wacker, W.: What's Your Story?: Storytelling to Move Markets, Audiences, People, and Brands, p. 39. FT Press, Upper Saddle River (2008)

McCulloch, S.: Engaging with the hard-to-reach. Good Governance Institute (2020). https://www.good-governance.org.uk/publications/insights/engaging-with-the-hard-to-reach#:~:text=Hard%2Dto%2Dreach%20groups%20tend,aren%27t%20heard%20or%20considered

Norris, P.: Social Inequalities. Digital Divide: Civic Engagement, Information Poverty, and the Internet Worldwide. Communication, Society and Politics. Cambridge University Press, Cambridge, pp. 68–92 (2001)

Pappers, J., Keserü, I., Macharis, C.: Co-creation or public participation 2.0? An assessment of co-creation in transport and mobility research. In: Müller, B., Meyer, G. (eds.) Towards User-Centric Transport in Europe 2. LNM, pp. 3–15. Springer, Cham (2020). https://doi.org/10.1007/978-3-030-38028-1_1

Puumala, E., Leino, H.: What can co-creation do for the citizens? Applying co-creation for the promotion of participation in cities. Environ. Plan. C Polit. Space **39**(4), 782–784 (2020)

Royo, B.: D3.1- INDIMO Pilots handbook. INDIMO project deliverable (2020)

Saha, G.G.: A paradigm shift from digital divide to digital inclusiveness. IBMRD's J. Manag. Res. **3**(1), 75–84 (2016)

Tovaas, K., Rupprecht Consult: How to make inclusive mobility a reality: 8 principles and tools for a fair(er) transport system (2020)

Vanobberghen, W., et al.: D1.2 - User needs and requirements on a digital transport system. INDIMO project deliverable (2021)

126   K. Bulanowski et al.

INDIMO Communities of Practice in Monghidoro, Antwerp, Galilée, Madrid, and Berlin: A Common Space for Co-designing Inclusive Digital Mobility Solutions

Floridea Di Ciommo[1]([✉]), Eleonora Tu[2], Juanita Devis[3], Michelle Specktor[4], Yoram Shiftan[5], Miguel Jaenike[6], Gianni Rondinella[1], Esau Acosta[6], Thais Lamoza[7], Martina Schuss[8], and Arne Nys[7]

[1] cambiaMO | changing Mobility, Madrid, Spain
floridea.diciommo@cambiamo.net
[2] Fondazione ITL, Bologna, Italy
[3] Vrije Universiteit Brussels, Brussels, Belgium
[4] The Israeli Center for Smart Transportation Research, Technion, Israel Institute of Technology, Haifa, Israel
[5] Civil and Environmental Engineering, Urban and Regional Planning, The Israeli Center for Smart Transportation Research, Technion, Israel Institute of Technology, Haifa, Israel
[6] Vivero de Iniciativas Ciudadanas (VIC), Madrid, Spain
[7] door2door GmbH, Berlin, Germany
[8] Technical University Ingolstadt, Ingolstadt, Germany

Abstract. This paper demonstrates the co-creation process of digital mobility and delivery services applied in the Inclusive Digital Mobility Solutions (INDIMO) project mainly based on the local Communities of Practice (CoP) drawing on the knowledge and experience of their members to propose solutions adapted to their needs and interests. In the context of the INDIMO project, CoPs were established at five pilot locations and included users, mobility service providers, (digital) developers, user interface designers, and policymakers associated with each pilot. This chapter aims to report on the experience of the INDIMO project in employing the CoP as a tool to integrate the development of digital mobility and delivery services and the contribution and cooperation of different actors such as operators, developers, policymakers, and organizations representing the end-users. The creation of common spaces such as the INDIMO communities of practice was fundamental to enhance cooperation among different actors, co-design inclusive digital mobility solutions, and empower the participants in using the above-mentioned services. This chapter shows the development of the CoP process, the activities and challenges, and its role in making digital mobility services inclusive and universally usable.

1 Introduction

Communities of Practices (CoP) are a group of people who share a concern or a passion for something they do, and learn how to do it better as they interact regularly (Wenger

© The Author(s) 2023
I. Keseru and A. Randhahn (Eds.): *Towards User-Centric Transport in Europe 3*, LNMOB, pp. 127–141, 2023.
https://doi.org/10.1007/978-3-031-26155-8_8

et al. 2002). CoPs have been extensively observed in a variety of contexts, such as in the educational or medical sector. The Inclusive Digital Mobility Solutions (INDIMO) project has explored the potential of Communities of Practice (hereinafter referred to as CoPs) by setting up and running local CoPs for its pilots where a digital mobility or delivery service was developed or redeveloped taking the principles of Universal Design into account. Each Community of Practice brought together local users, mobility, and delivery service providers, (digital) developers, user interface designers, and policy-makers associated to each pilot. At the beginning of the project, for each pilot, a call for participation was launched to engage interested parties into a collective learning process. The main objective of these Communities of Practices was to contribute to develop the INDIMO Co-creation Community and to identify the profiles of user groups in situation of vulnerability with respect to the digitalization of mobility with their requirements and needs.

The co-produced common knowledge within the local CoPs was consolidated and served as structured feedback for developing various components of the INDIMO digital mobility toolbox, including the Universal Design Manual (UDM), the Universal Interface Language (UIL), the Cybersecurity and privacy assessment guidelines (CSG), the Service evaluation tool (SET) and designing social and educational strategies for enhancing the appropriation of the use of digital mobility services (DMS) and digital delivery services (DDS).

The creation of a safe space such as the INDIMO CoP was crucial to build up common knowledge on the local context, inclusive digital mobility solutions and empowering the participants, including the target groups of vulnerable users. We concretely learned that the creation of the shared space provided by the Communities of Practice (CoP) during the project timeline enhanced the collaboration among different actors who oversee digital, physical, and regulatory features of digital mobility and delivery solutions. In the next sections, we will outline the setting up process of the local CoPs' and their role in identifying the key elements that digital mobility and delivery services should include to be inclusive and universally usable. Section 2 includes the description of the key characteristics of the five Communities of Practice located in each INDIMO pilot location. Section 3 presents the typology of members for each CoP, while Sect. 4 underlines the notable points of the CoP activities and the results of each CoP conversation. In Sect. 5 we share the lessons learned. Finally, insights and conclusions are summed up in Sect. 6, including a reflection on the relevance of sharing different knowledge and perspectives with a universal design approach in the era of digitalization.

The results presented in this chapter refer to the 3 years when the INDIMO project was developed (2020, 2021 and 2022). The content is mainly based on the INDIMO Deliverable D3.2 *Communities of Practice Report*.

2   The INDIMO Pilots and Organization of CoPs

Each pilot and the related CoP in the INDIMO project had a different aim with the common goal of improving the inclusion and accessibility of digital mobility services for the identified target groups of people in situation of vulnerability (see Fig. 1). Local CoP members participated in the co-creation process of the five pilots. As an open group, anyone could join and leave at any time, yet CoPs were made up of people who were interested in the development and potential growth of the different digital mobility and delivery services.

The figure below shows an infographics of the five different INDIMO pilots, indicating in the outer circle the digital mobility or delivery service of the pilots and in the inner one the target groups that were involved.

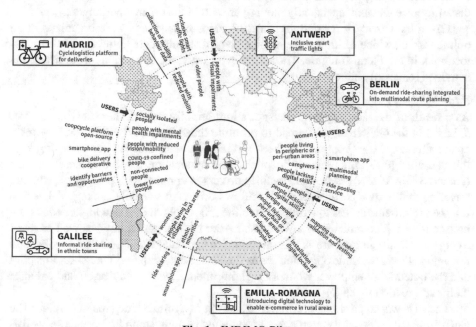

Fig. 1. INDIMO Pilots

Pilot 1-Monghidoro, Italy

Pilot 1 (P1) was located in Monghidoro (Emilia-Romagna Region) and aimed to introduce digital technology to enable e-commerce in rural areas to increase the inclusion and accessibility of digital mobility services. Its main activity was the installation of an advanced digital locker operated by Poste Italiane. This service offers the possibility to ship and collect parcels, collection of signed correspondence, postal bill payment, recharging of telephone cards and prepaid Poste-pay cards, and management of customer-to-customer deliveries. The main target population group of P1 were rural residents of Monghidoro, specifically older people with a low level of digital skills and low-income migrants with lower proficiency of the local language (i.e., Italian).

Monghidoro is a small municipality with about 3700 inhabitants about 50 km from Bologna. Based on the index of the Metropolitan City of Bologna, it is classified as highly vulnerable because of a combination of reasons such as low income, a high number of elderly people and migrants, and decreasing population.

Due to the Covid-19 restrictions, the CoP meetings in this pilot were organized both in person and online, participants had a preference for holding in = person meetings.

Pilot 2-Antwerp, Belgium

Pilot 2 (P2) based in Antwerp set up a CoP focusing on the inclusive smart traffic lights and had the purpose of sharing experiences and providing feedback about the proof-of-concept smart traffic light smartphone application, physical and online accessibility, the use of the public space, and the relationship of pedestrians with other modes of transport (i.e., cars, bikes, e-scooters). The CoP was a space for sharing knowledge and collectively solving problems about specific conditions that older people with visual disability face when independently moving around in the urban environment.

Given the Covid-19 restrictions, most of the CoP meetings were organized in an online format, which facilitated the engagement of stakeholders living in very different locations in the Flemish region. Therefore, even when the restrictions were lifted, the activities remained online.

Pilot 3 (P3)-Galilee, Israel

As a result of the lack of public transportation, an informal rideshare service in Pilot 3, based in the Galilee region, aimed to promote the status and rights of Arab women, create alternative travel options for them and increase their mobility for achieving key-life activities (e.g. work, school, hospital). The application for the rideshare service (named SAFARCON)[1] is free and enables drivers to connect with passengers who are traveling to the same destination. It also allows drivers to schedule package deliveries. It aimed to provide access to workplaces in the metropolitan area and higher education both to the Arab community as a whole and Arab women particularly, as well as those living specifically in rural areas. It is important to understand that in Galilee, women do not enjoy always the same opportunities as their male counterparts. Therefore, this pilot had the potential to empower women by giving them access to education and enhance their participation in the job market.

The idea was to promote informal ride sharing in a rural ethnic area of Galilee. The CoP discussed accessibility approaches to an inclusive digital mobility service, as well as the consideration of both gender and ethnic perspectives, language and cultural barriers to service access, and accessibility to technology for vulnerable minority groups of Arab women.

As part of periodic open group CoP meetings, a wide range of participants had the chance to share their perspectives, knowledge, and experience, including users, mobility service providers, digital application developers, user experience designers, and policymakers involved in the pilot, as well as other stakeholders including traditional transportation and digital mobility actors.

The collaboration and interactions between participants were fundamental in enhancing inclusive digital mobility solutions, facilitating the progress of the Galilee Pilot, and advancing INDIMO as a whole.

Pilot 4 (P4) -*Madrid, Spain*

---

[1] Technion's Transportation Research Institute (TRI) developed the application SAFARCON in collaboration with Kayan, a feminist Arab non-profit organization, under a grant from the Office of the Chief Scientist at the Israeli Ministry of Transportation (MoT) several years ago.

Pilot 4 was located in Madrid and aimed to enable food delivery for people with impairments, such as reduced mobility or vision, mental disability, low-income people, lower digital connection, socially isolated (e.g., unwanted loneliness), Covid-19 isolation through a digital application. The use of an existing cycle logistics platform was tested from users' experience and needs perspective to make this platform more inclusive and accessible. The digital food delivery platform was developed by the European Federation of Bicycle Delivery Cooperatives (Coopcycle), based in France, and it enables cooperatives of riders to operate a food delivery service at the local level with an inclusive approach. In Madrid, the service is operated by La Pájara, which operates the food delivery service from a set of more than 30 restaurants, including vegan ones, to users via the Coopcycle website and application. The main aim of this pilot was to improve the access of the target groups of the population to healthy food.

The CoP in Madrid aimed to discuss among the relevant actors the proximity and ease of use of food delivery as a way of tackling the isolation of target populations. Safety and security concerns were also discussed, along with the specific situations that arose during the Covid-19 pandemic. Digital skills and technological barriers were also important matters of discussion.

The CoP was a safe space for sharing knowledge, while discussing accessibility and inclusivity of a digital delivery service in an urban context. The CoP was intended to find solutions to accessibility problems for specific target groups of the population. Given the Covid-19 restrictions, also this CoP meetings were mainly organized in an online format. This format facilitated the engagement of stakeholders living in very different locations in Madrid. Even when the restrictions were lifted, the activities remained mainly online with some events in person.

Pilot 5 (*P5*)-*Berlin, Germany*

Pilot 5, based in Berlin, had the initial aim of increasing access and providing individual mobility for women as caregivers though a door-to-door ridesharing service. The service sought to improve short-distance mobility of women with children, offering a connection to public transport stations, facilitate short walking distance to pick-up and drop-off points, cover the lack of transport services in peripheral locations and lack of digital skills or low speed of internet connections. An existing multimodal mobility platform was tested by users to make this platform more inclusive and accessible to women as caregivers. This digital platform for trips planning had been developed by the IT developer of door2door and it enables public transport and digital mobility operators to provide an integrated service (i.e., MaaS).

The CoP focused on co-creating knowledge, discussing mobility needs of women as caregivers and sharing their experience in public transport and space. It was mainly oriented to the co-design of digital mobility services that generate confident relationships among caregivers and women, empowering them and fostering sharing of experiences through empathy.

3   INDIMO Community of Practices: Activities, Characteristics and Participants

This section will present the activities, characteristics, and participants of the five CoPs set up within the INDIMO project. As the target groups varied greatly among the pilots, participants also varied among the different CoPs and over time within each CoP. However, the aims remained focused on giving participants a platform to share their life experiences for improving a human-centered design of digital mobility solutions.

The CoP of Monghidoro (P1) included institutional actors such as the data and ICT services provided for Emilia-Romagna Region (LEPIDA), the Municipality of Monghidoro with its Mayor; the Metropolitan City of Bologna with its Mobility planning and social area department, rural residents, and two local NGOs, one representing elderly users (Le Pozze) and another one representing especially migrants' residents. This CoP was specifically oriented to involve other rural residents coming from municipalities close to Monghidoro, elderly residents, and residents of migrant descent of non-Italian nationality. It was facilitated by INDIMO partners, including the Institute for Transport and Logistics, Poste Italiane, cambiaMO, and DeepBlue.

The Antwerp CoP (P2) community was composed of a diverse group of stakeholders, representing different perspectives and roles in the digital mobility ecosystem. The Community members consisted of intersectional representatives of elderly people and people with visual impairments, developers, and user interface designers, policymakers from the Antwerp municipal services of mobility, regional accessibility agency, mobility service operators, and, most importantly, end-users in situation of vulnerability.

Facilitated by the online format, the Antwerp CoP was able to establish a broader stakeholder community and expand its scope and impact from the city of Antwerp to the Flemish region. In this way, the CoP was able to engage users, representatives, civil servants, and other stakeholders in providing their views coming from the perspectives of the different cities. Moreover, it facilitated the identification of common regional challenges and needs.

To gather feedback from visually impaired users, in some CoP activities, we conducted a series of one-to-one interviews. The interviews followed the same format as the CoP activity. The outcomes of the interviews were then shared during the CoP with other stakeholders. In some situations, where the one-to-one interview was not possible, a breakout room with a facilitator was created for each visually impaired participant.

The CoP in Antwerp (as most of the other CoPs) set up the invitations with a clear agenda and expectations for the different stakeholders. This proved to be effective in maintaining the active participation of the different stakeholders throughout the INDIMO CoP cycle. The type of stakeholders participating in each activity, therefore, varied according to the topic of the activity. As an example, in the activities related to the end-user appropriation of digital mobility tools or the re-design of the proof-of-concept application the feedback from User Interface designers, developers, and visually impaired users proved to be more relevant. Moreover, we learned that stakeholders, especially civil servants, and service providers, are more eager to participate if there is a dissemination of the results during the CoP activities.

The CoP of in Pilot 3, in Galilee was the most diverse in terms of its members during the timeline of its meeting sessions. Given this diversity, the authors of this chapter found it appropriate to list the members in Table 1. All CoP members were invited to join CoP meetings regularly, yet each meeting consisted of a slightly different mix of seasoned participants making each meeting dynamic and interesting in its way.

In the pilot 4 in Madrid, we organised regular meetings and included a wide range of participants, such as:

- policy maker from the Municipality of Madrid,
- representatives of target groups including the *CEAPAT- Imserso*, representing persons with both mental and physical impairments and elderly,
- *Tangente,* representing the elderly women socially isolated,
- *Asidown*, representing persons with a mental disability such as down syndrome
- *ONCE*, representing persons with reduced vision,
- Representatives from the delivery- operator La Pajara,
- Digital developers (from Coopcycle, the cooperative digital platform used by la Pajara, and another representing as well the end-user person with reduced vision)
- User Interface designers

These participants had the chance to share their points of view, knowledge, and experiences from different perspectives: as users, mobility service providers, digital application developers, user experience designers, and policymakers. At the beginning of each session, participants introduced themselves and their motivation for taking part in the CoP. The CoPs activities included the assets, values, and the functioning of the platform; the users' requirements prioritization, the persona construction; and the co-design of a mobility and delivery inclusive service.

Finally, for pilot 5 in Berlin, the Community of Practice mainly included (digital) developers and User Interface designers, product managers, a researcher in gendered mobility, and public transport representatives including local users, mobility and delivery services providers, to develop common knowledge around the inclusiveness of the ridesharing app.

4   Relevant Topics and Issues that Emerged During CoP Activities

The main topics that arose during the CoPs co-creation and co-design processes include different key themes:

1) the discussion around capabilities, limitations, and requirements elaborated through the analysis of semi-structured interviews,
2) the rating exercise of the users' requirements that have nourished the elaboration of the Universal Design Manual,
3) the assessment of the icons used in the various digital services for creating the Icons and Languages catalog, and
4) the co-design of the implementation of the recommendations for an inclusive and accessible digital mobility and delivery service.

Table 1. P3-Galilee community members

#	Title	Type
1	Israel's National Accessibility Supervisor Equal Rights Commission for the Disabled, Ministry of Justice	Policymaker
2	Chief Scientist at the Israeli Ministry of Transportation	Policymaker
3	Former Treasurer at the Israeli Ministry of Transportation involved with SAFARCON App original development	Policymaker
4	City Planner, The City of Jerusalem	Policymaker
5	Strategic planning at Ayalon Highways	Policymaker
6	World Bank Transportation Strategy	Policymaker
7	http://www.thejoint.org.il - The Joint - independent living for people with disabilities	NGO, influencing Policy making
8	http://www.techpolicy.org.il - Israel Tech Policy Institute	NGO, influencing Policy making
9	http://www.Nanooa.org.il - promoting realization of opportunities inherent in the field of smart-transportation for the benefit of the public	NGO
10	https://www.kayanfeminist.org/home-page Kayan - an Arab feminist organization with the goal of advancing the status of Arab women in Israel and protecting their rights	NGO. Users' representative
11	Arab women - SAFARCON drivers and passengers	Users
12	Arab women students and researchers	Users' representatives
13	Nadsoft – Arabic-speaking software developers	Developer
14	Urban Mobility software developer	Developer
15	Researcher in Human Factors Design	UX/UI
16	Cactus Ads and More – Arabic-speaking digital advertising and marketing office	Marketing
17	Transportation planner involved in SAFARCON development	Planner
18	Head, The Israeli Smart Transportation Research Center	Researcher
19	Transportation researcher Involved in SAFARCON development	Researcher
20	Researcher in Ride Sharing	Researcher

(*continued*)

Table 1. (*continued*)

#	Title	Type
21	Researcher Travel Behavior and Safety Evaluation	Researcher
22	Researcher Candidate, Smart Mobility, and the Human Factor	Researcher
23	INDIMO Partners	Researcher

In this section, the main topics from above are presented. It is, however, very important to underline that CoPs faced very practical problems, such as the difficulty of organizing the events at a time suitable for everyone, as policymakers are generally available during the day while users or associations can participate in the CoP activities outside working hours.

In the case of pilot 1 in Monghidoro, one issue was related to the difficulties of meeting on-line. As the target groups were elderly people and people with low digital skills, it was challenging to involve them using the virtual space. Elderly people were involved but health issues related to Covid-19 complications meant that they were no longer able to join. Despite these difficulties, the CoP meetings were lively and explored various challenges, and several themes emerged concerning the domain of the smart locker service implementation in a rural context. These themes are explained below.

During the 11 CoP sessions, there were concerns with regards to the deploying of the service of a digital locker, and particularly if this was meant to replace the existing physical Post Office. The digital locker was perceived as a threat, as in the eyes of the participants it could overtake the role of the physical office. In addition, in the small municipality of Monghidoro, rural residents recognized and attributed social value provided by the few (public) meeting points (such as the post office).

The negative perception of the digital locker is probably connected to the fact that over time, Monghidoro residents had perceived a decrease in the accessibility and availability of basic services. Some activities of the CoP were, therefore, also focused on enabling participants to understand that the digital locker service was an additional service and not a replacement for the current national postal service.

In the case of pilot 2 in Antwerp, the most important topics discussed during the 13 CoPs included: (i) the importance of a well-designed and accessible urban environment, (ii) the different needs of visually impaired users concerning other types of vulnerable populations, (iii) the importance of peers in adopting an application, and (iv) the recurrent issues visually impaired users experience when using digital tools. The complexity and unpredictability of the urban built environment was a key limitation in the autonomy of visually impaired users. As an example, ramps on the sidewalks to cross the street were perceived as a physical barrier for visually impaired people since their dogs could not distinguish where the road started.

Technology has a key role in supporting visually impaired people in navigating the physical space and is evolving to incorporate these urban elements. Visually impaired people tend to be very skilled in using digital tools, older people instead experience many challenges related to technology adoption and usage. Older people, even if affected by visual limitations, have very different needs and technical skills than the average visually impaired people. Today, readers incorporated into different digital systems enabled the usage of any application for visually impaired users. However, not all the applications are accessible to them, making the process of selecting and downloading the apps is difficult. Even more, the reviews of the applications do not take into consideration accessibility for visually impaired users. Consequently, the adoption process relies mostly on the recommendations of their peers.

In pilot 3, in Galilee, CoP meetings were organized around well-defined agendas that included icebreakers, polls, and role plays followed by an open discussion. As a first step, it was important to have the developer explain SAFARCON, the app that aimed to address mobility needs of the Arab community and women. One of its main goals was to improve its service accessibility. CoP participant stakeholders noted that mainly young people downloaded the application and brought up the issue of women's experience. Women must feel comfortable ridesharing with strangers and be confident about the safety of the vehicle. App integration with social media was highlighted as a safety measure, as users would be able to know their Facebook friends' usage and opinions.

The CoPs discussions revolved around how social media influencers, radio shows, and authorities could endorse and promote app use. The main conclusion was that it was crucial to know who the recipients of the design were, because icons do not have a single meaning[2].

Looking at how social norms, culture, language, and gender influence the appropriation of mobility apps, we found that it boils down to subjective personal preference, taking into consideration usability, privacy, security, accuracy, reliability, and functionality. In addition, limitations in storage or battery usage of apps, as well as cyber security concerns, affect the adoption and use of digital mobility apps[3].

The outcomes of the activities developed within the CoP in pilot 4 in Madrid were beneficial over a wide range of aspects, from the fieldwork deployment of tasks related to the specific objectives of the INDIMO project to the co-created implementation of service improvements and organizational measures. Conducting the co-creation work implied a good number of challenges and obstacles that had to be overcome, especially when the aspiration was to build up a local Community of Practice among stakeholders and researchers who have not experienced this kind of knowledge consolidation tool. The main ideas of the CoP were oriented towards the service and the social economy and sustainable values that it conveys, the actual and potential target audience, and elements of usability of the app itself.

Another argument that emerged was the focus on the ethical approach of the delivery company, mainly towards its workers, and in its way to offer a more inclusive and accessible digital delivery service. This last point implied the recognition of a limitation that

---

[2] Stakeholders' suggestions to make the app more accessible and help overcome digital barriers were summarized in the INDIMO Deliverable D2.1 - Universal Design Manual and the detailed assessment of screens and icons is presented in the Deliverable D2.3 corresponding to the Universal Icons Language.

[3] The INDIMO Deliverable D2.5 Enhancing Appropriation of Digital Mobility Solutions contains a more detailed analysis of the results of the appropriation exercise.

La Pájara had: a very homogenous base of users, characterized by their young age, high level of education, digital competency, and social and environmentally aware population. Nevertheless, the type of inclusive digital and physical services adaptations and how to implement them were also raised during the CoPs debates. There is a realization that attaining inclusivity in the digital realm takes effort. Full comprehension of the concept of universal design appears.

Participants understood and suggested that the co-design and update of the delivery service platform goes beyond simply adjusting for special needs, but rather aims to meet a wide range of universal requirements. There were some people who were concerned about the fact that the app was too focused on food delivery, and not on a more general courier service, from which it could benefit. The emphasis was on organic growth and inclusion of new users, especially coming from the target-groups of people in situation of vulnerability that are not always addressed by other food delivery platforms, recognizing that food delivery can facilitate their daily lives. During the debates, some people also suggested micro-training for riders, in order to better address the needs of people in situations of vulnerability. In the third phase of the pilot, (deployment), this suggestion was realized through the inclusive training for riders.

CoP meetings of pilot 5 in Berlin were systematically organized around an agenda related to the co-creation activities of the INDIMO project oriented to identify the capabilities and requirements of women caregivers and with children; elaborate and select the Universal Design Manual recommendations for making the ridesharing service more inclusive for women caregivers of children and finally redesign the service by prototyping an app-based security function including three different degrees of safety: the emergency assistance bottom, silent alert for the driver, and silent alert driver's view. During the first meetings, the participants were asked about their interests and how they be could integrated into the discussions of the community of practice. There were doubts among participants about the project, whether it is about the development of an application or a local service strategy. Participants introduced themselves and described how they could contribute to the pilot. There were a few main points raised during the debate on the timeline. The mobility of women with children is a particular concern. One participant claimed that a similar service (titled Berlkönig) was offered as a ride-pooling service in a different peri-urban area. The importance of understanding the needs of women as caregivers was emphasized, specifically since transportation organizations rarely involve women in their discussions.

5 Outcomes

In this section the outcomes from the individual CoPs in the five pilots are summarized.

For pilot 1, in Monghidoro, the matter of the digital divide due to age and the adaptation and interest of different generations to the technology is a subject that frequently arose. It is expected that the impact of the inclusion will be differentiated for different segments of populations. Many residents (not just elderly people) feel unfamiliar with services that rely on "technology", intended as mobile phone apps, QR codes, payments through the internet, and so on. The need for on-site training, of a person, that "can show how it is done" and the initial availability of in-person assistance must be considered to introduce rural elderly and lower-skilled target groups of users to this digital service without fear. This need was also recognized as the top priority for policymakers, end-users, and stakeholders' organizations across all pilot processes.

The outcomes of the CoP in pilot 2 in Antwerp served to improve the different aspects of the CoP itself, from the organizational part to the way each CoP content was delivered to the participants. The main learnings for P2 included the following recommendations:

- Send a clear agenda with a clear description of what the expectations are towards each type of stakeholder. This would allow CoP participants to decide if their participation is relevant.
- Dedicate some time of the CoP to share project progress and learnings. Stakeholders, especially civil servants, and service providers will be more eager to participate.
- Check the accessibility of the tools and materials you will use during the CoPs.
- Engage end users in providing feedback about the use of digital tools. Representatives do not have deep knowledge about the use of digital mobility applications as visually impaired users.
- Organize one-to-one interviews or breakout rooms to facilitate the collection of feedback from visually impaired users. Visually impaired users are overwhelmed in large meetings. They often have difficulties following and participating in activities.
- Send the material in advance to visually impaired users. In this way, they will be able to read the questions and reflect on them before the activity.

In the case of the pilot 3 in Galilee, the expectation to build a local Community of Practice including stakeholders and researchers was challenging. Many of these stakeholders had limited time and no prior experience with such a method. This posed challenges that had to be overcome during the co-creation process. Despite this, as soon as they accepted the invitation to participate, they were engaged and wanted to contribute with their knowledge and insights. Additionally, the open group allowed participants to join and leave at their convenience, making it easier for them to commit according to their schedules.

Fundamentally, it is crucial to think of "Accessibility" from the perspectives of both "App Accessibility" (digital accessibility) and "Service Accessibility". An app may be accessible, but the service itself may not be, for example, a vehicle that is too high for a person with a disability.

Local CoPs played a key role in identifying and recommending the key elements that a mobile and delivery service should contain to be inclusive and universally usable. The need for human assistance was the top priority for policymakers, end-users, and stakeholders' organizations across all co-creation activities relating to inclusive digital mobility.

At the time of Covid-19, the CoP had to be carried out on a digital platform. Some participants had very low familiarity with any type of digital tools and in some cases, for older people, it was the first time that they participated in a videoconference. The availability of the local pilot leaders was the key to guarantee the success of CoPs meetings. This is a clear example of empowering target-groups, that makes the CoP unique in the co-creation process. As a result, CoPs had resources for hypotheses, conjectures, and possible scenarios for participants to talk about their beliefs and feelings about technology and mobility and delivery services.

Users sometimes they talked freely about their concerns and interests in general digital services or general characteristics of technology in services. Thus, the facilitator should drive them back into focus on the CoP aims. There is a general feeling of having consolidated local CoPs in a good way. The great number of verbatims and contents produced for the rest of INDIMO tools (i.e., Universal Design Manual and the Universal Icons languages catalog) anticipates a high level of inputs for clear guidelines for the INDIMO Digital Mobility toolbox.

The recursive appearance of beliefs, motivations, and feelings shared by several participants lead to believe that there are social representations and images about digital mobility and delivery services that should be considered at the time of designing technology for including end-user target groups.

The recruitment of end-users was a challenge for pilot 5 in Berlin because the local partner Door2Door was a white-label platform offering the integrated ride-pooling service to public transport operators and mobility providers but not operating its own service. Then, single mothers in Berlin, Germany were recruited through the Door2Door network responding to profiles through fieldwork in Marzan, a suburban municipality where this profile of person in situation of vulnerability was concentrated. Concretely, the CoP of Berlin shows the development of an emergency button concept that focuses on perceived safety. Therefore, different scenarios were derived to understand when women feel uncomfortable when using ride-pooling services. Three stages of severity of perceived safety are conceptualized: yellow (relatively safe), orange (moderately safe), and red (extremely unsafe).

Subsequently, developers in Berlin, prototyped an app-based security function for each of these situations. This concept goes beyond the well-known emergency breaks in public transportation systems and takes into consideration that security-critical situations many times are very nuanced and need differentiated solutions.

6 Lessons Learnt Across the Pilots

The INDIMO CoPs revealed several common aspects. Firstly, conducting the co-creation work implied a good number of challenges and obstacles, especially when there was the aspiration to build-up a local Community of Practice within stakeholders and researchers who did not have experience with this kind of knowledge consolidation tool. To support this process, bilateral and collective training sessions were arranged with pilot leaders before starting the meetings of the CoP.

Secondly, all CoPs co-created and contributed with valuable inputs to the INDIMO toolbox within local stakeholders and target-groups of users. Examples of the contribution were the following: prioritizing the requirements to make digital mobility and delivery services inclusive, selecting the appropriate interface icons for different scopes, insights on the appropriation digital tools for carrying out daily live activities of people in situations of vulnerability.

Thirdly, the conversations within the COPs were mainly focused on the problems and participants were eager to contribute and to find common solutions to these common problems. This feeling arose from the CoP where all the practitioners were enthusiastic about their participation and felt safe and comfortable in the space (both virtual and physical) created to talk about these issues. Most participants were aware of the benefits and disadvantages in developing inclusive digital services.

6.1 New Insights and Conclusions

Despite their differences the CoPs provided several interesting general outcomes. In this section, the key insights and conclusions are presented. These are common across all CoPs.

CoPs can be used in a variety of contexts, and specifically in the realm of digital mobility services, where they can provide expertise and knowledge coming from specific target groups.

There is a need for training facilitators when CoP are set up and run. This training is essential for organizing the CoPs as a safe space for co-creational and co-designing activities. In our case, the profile of a CoP facilitator featured technical notions of Digital Mobility and Delivery services, as well as (and perhaps more importantly) excellent communication and listening skills.

The COVID-19 pandemic measures imposed the use of virtual space for running the INDIMO CoPs. This aspect was a real challenge for involving our user target groups who are per se quite digitally low-skilled. Nevertheless, the digital space was the opportunity to make the digital session more dynamic and provide a pleasant space where people would stay and have a good moment to share their experiences and knowledge. These COPs outcomes were achieved thanks to the previous experience of the COPs task leaders in managing complex and conflictive co-creational processes within the COPs methodological approach. In general, the digital sessions have been run in all kinds of platforms: GoToMeeting, Teams, Zoom and Google Meets. All of them work quite well with light preference for the Zoom virtual space that was more user-friendly for less digitalized participants.

Mutual support plays a key role in this INDIMO CoPs co-creation process. It was crucial to establish relationships based on trust, both within the CoPs but also among the 5 CoPs and with the collective training sessions provided by cambiaMO and VIC, both INDIMO partners.

CoPs were effective to produce knowledge and prioritize recommendations for the design and deployment of inclusive digital mobility services. Outcomes from the CoPs were also the input to several tools that the INDIMO project developed (e.g. toolbox of the Universal Design Manual - UDM, Universal Icons Languages-UIL, the cyber security and privacy assessment guidelines -CSG, and the Stakeholders evaluation tool – SET).

Acknowledgement. The research is a part of the Inclusive Digital Mobility Solutions (INDIMO), a Horizon 2020 project that has received funding from European Union's Horizon 2020 research and innovation programme under grant agreement no. 875533.

References

Wenger, E., McDermott, R., Snyder, W.: Cultivating Communities of Practice. Harvard Business School Press, Boston (2002)

Di Ciommo, F., Rondinella, G., Kilstein, A.: Users capabilities and requirements. Deliverable D1.3, INDIMO (2021a)

Di Ciommo, F., Kilstein, A., Rondinella, G.: Communities of Practice, Deliverable D3.2, INDIMO (2021b)

Di Ciommo, F., Kilstein, A., Rondinella, G.: Universal Design Manual Deliverable D2.1, INDIMO (2021c)

Hueting, R., Giorgi, S., Capaccioli, A., Bánfi, M., Soltész, T.D.: Universal Icons Language Deliverable D2.3, INDIMO (2021)

Specktor, M., et al.: Enhancing Appropriation of Digital Mobility Solutions, Deliverable D2.5, INDIMO (2021)

Mainstreaming Sustainable Urban Mobility – The Mieri-Mobil Project

Michael Abraham[1](✉) and Carolin Schröder[2]

[1] Kommunales Mobilitätsmanagement, Deutsche Plattform für Mobilitätsmanagement (DEPOMM e.V.), Berlin, Germany
info@michaelabraham.de
[2] Zentrum Technik und Gesellschaft/Technische Universität Berlin, Berlin, Germany
c.schroeder@ztg.tu-berlin.de

Abstract. In 2021, the Berlin district Charlottenburg-Wilmersdorf offered residents various mobility options on a part of an urban square called Mierendorffplatz. The overall objective was to explore how people living at the square or in the closer neighbourhood could change their mobility habits to become more environmentally friendly. The test aimed at exploring how sustainable urban mobility can be successfully implemented on a wider scale in in other parts of the city - thus seeking to contribute to mainstreaming sustainable mobility planning in regular city-wide mobility planning processes.

This article describes results from evaluating the project: Implementing such exploratory living labs successfully and without excluding specific user groups, depends on the integration of knowledge and resources of both the local governing bodies and civil society. This puts a focus on the question of accessibility of urban space and on the need to re-organize it differently if sustainability criteria should be met.

It concludes that sustainable forms of mobility need to become more demand-orientated and that different legal and economic frameworks are necessary to make it a real alternative for everybody. These two aspects will be elaborated. The findings are based on empiric evaluation and validated through consultations with high-ranking experts.

1 Introduction

From mid-June to the end of December 2021, the Berlin district of Charlottenburg-Wilmersdorf implemented the project Mieri-Mobil on the eastern part of southern Mierendorffplatz (www.mieri-mobil.berlin). Located in the central part of Berlin, it is connected to the public transportation system by bus, subway and city train (S-Bahn). For many years, this district has not been affected too much by gentrifications processes - a large number of residents lives there for decades. But in 2017, the demographically largest group were people in work between 25 and 45 years as well as families with children not attending primary school yet (BA Charlottenburg-Wilmersdorf 2017).

The project aimed at offering different forms of alternative mobility to exactly this mix of local residents. This was based on the basic assumption that alternative forms

I. Keseru and A. Randhahn (Eds.): *Towards User-Centric Transport in Europe 3*, LNMOB, pp. 142–153, 2023.
https://doi.org/10.1007/978-3-031-26155-8_9

of local mobility are more sustainable in both environmental and social aspects than privately owned cars. Another aim of the project was to carefully evaluate the project. The approach pursued was to not only assess the actual effects of alternative mobility services and forms but also the planning and implementing procedures that accompanied the project. This way it should be made sure that factors leading to a successful implementation and transferability criteria could be identified.

The legal basis of the planned project is Section 45 of the German Road Traffic Act "§45 traffic signs and traffic facilities.

The road traffic authorities may restrict or prohibit the use of certain roads or stretches of road for reasons of safety or traffic order and divert traffic. They have the same right… 6. to investigate accidents, traffic behaviour, traffic flows and to test planned traffic safety or traffic regulation measures"[1].

The Mieri-Mobil project was characterized by three unique aspects: First, it focussed on introducing new forms of sustainable and smart mobility. Second, the project was based on several mobility and logistics projects in the same area and with a similar group of actors that have been implemented since 2016 and their findings: these are NEW MOBILITY BERLIN, distribut-e, Stadtquartier 4.1, the KIEZBOTE and a mobility transformation concept for the Mierendorffinsel based on SUMP (Sustainable Urban Mobility Plan, Rupprecht Consult 2019). Third, it was evaluated by applying a specifically developed evaluation approach (by the authors of this article).

Over the entire period of the project, various forms of micro-mobility were offered in this one location. In addition to a mobility point (called Jelbi in Berlin) with rental bicycles, e-mopeds and e-scooters, the offers also included stationary and free-floating car sharing and cargo bikes. In order to create space for the new offers, existing parking spaces were temporarily rededicated to eight car-sharing parking spaces, a disabled parking space, a cargo bike parking space, a delivery parking space and a logistics hub for a local delivery service (Kiezbote), in the form of a container. Replacement parking spaces were provided for the residents about 200 m away.

Initially, at the beginning of planning the project it was also intended to set up a temporary play street. This linking of different approaches to put more attention to more sustainable mobility increased the experimental character of the project.

Time for preparing the implementation and the evaluation of the project was very short: the decision to implement the project was taken in November 2020 with the actual project work starting in January 2021. The inauguration was scheduled for mid-June 2021, leaving some five months to prepare the mobility offers and the evaluation. Informal discussions with residents and local business people started in April 2021, a formal information letter was sent in early May.

It should be noted that, contrary to the original planning, it was decided not to completely block the street as a temporary play street for car traffic for the duration of the campaign. This was mainly justified by a newly set up construction site on the other side of the square, which increased the parking space pressure in the surrounding area. Another reason for it were concerns of the market management, who feared delivery problems and a decline in customers due to the necessary facilities for the play street.

---

[1] Original text in German: http://www.verkehrsportal.de/stvo/stvo_45.php.

2 Evaluating the Project – Effects and Processes

Mieri-Mobil is a unique experiment in this specific location. As mentioned before, it combined a car-based shared mobility hub with additional features and aimed at providing analysis regarding the location and general idea of this area of sustainable and smart mobility. Mobility hubs, in their widest definition, are objects of a growing number of case studies and conceptual studies from all over the world (for example: Schelling 2021; Difu 2020; Bell 2019).

While research on mobility hubs is still an emerging topic, there are three early findings: First, the location should be selected carefully (van Gerrevink 2021; Bell 2019). Second, the provided services should be tailored to suit local needs and be integrated into the existing urban space and uses - and therefore may vary in size and offer (Arseneault 2022; Difu 2020). And third, that existing business models for mobility hubs and the provision of alternative mobility pose new challenges to local governance structures (Arseneault 2022; Pangbourne et al. 2018).

In this context, it needs to be stated that the idea of mobility hubs is still strongly rooted in the concept of mobility as a service (MAAS), focussing on the provision of new vehicles and infrastructures while actual user needs are not yet much in the focus of quantitative and qualitative scientific research (Bell 2019; Pangbourne et al. 2018; Utriainen and Pöllänen 2018).

It may be assumed that the private mobility providers do gain profound insight in the usage numbers and users of their services. But these data rather illustrate actual usages than real needs and are in most cases not accessible for scientific analysis for data-protection reasons (Hadachi et al 2018). Nonetheless, some approaches to analyse actual user demands from a qualitative perspective have been implemented. Methods used include on-site observations, random interviews and focus groups (Arseneault 2022; Bell 2019) as well as peer review workshops in order to frame and compare experiences with mobility hubs (Abraham et al. 2021).

With this in mind, the qualitative – and to a lesser extent the quantitative - evaluation of the implemented aims, measures and processes played an important role throughout the Mieri-Mobil project. The overarching goal of the evaluation was, on the one hand, the qualitative analysis of the use of the mobility offers and their effects on urban space (impact evaluation) and, on the other hand, the implementation of the project itself (process evaluation). Detailed results have been published in a technical report.

The impact evaluation recorded the effects of the implemented measures on two levels: Evaluation of the effects of individual measures and evaluation of the project as a whole. Methods used for impact evaluation were

- Participatory observation through regular on-site visits, including photographic documentation
- Qualitative survey of residents and local businesses on the use and location of offered mobility services in the period from August to December 2021, both on paper and online via the project website.
- Quantitative collection of selected data from mobility service providers
- Three resident workshops with participation of representatives of the district Council in June, August and September 2021

- A peer review workshop (focus group workshop) in February 2022

The process evaluation was performed on two levels as well: Evaluation of the processes of implementation individual measures and evaluation of the implementation processes within the whole project. Methods used for process evaluation were

- A process evaluation workshop on October 14, 2021 with representatives of the Center for Technology and Society of the TU Berlin (ZTG), the insel-projekt and Schröder&Abraham GbR
- Two resident workshops with the District City Council in August and September 2021
- A peer review workshop (focus group workshop) in February 2022

3   Results - Impact Evaluation

Impact evaluation comprised mainly counting the numbers of usages of the new mobility services, a survey among neighbors of the street and the closer surrounding of the square, and information gained by observing how people are dealing with the new services on site. The most relevant and exemplary results are summarized in the following.

User numbers provided by the operators of the new mobility services: After the new mobility service was introduced in June, the number of Lime Scooters rented at Mierendorffplatz rose steadily to almost 70 rentals in August. Thereafter, the number dropped sharply until December, presumably due to deteriorating weather conditions.

The number of MILES rental cars rented at Mierendorffplatz increased after the introduction to more than 50 car rentals in November. Then, the number dropped to 35 by January 2022. The reasons for this are probably the high number of public holidays in December and early January, the reintroduced lock-down caused by the Covid pandemic, or fewer vehicles available for hiring. A correlation of the number of uses with bad weather could not be observed for the closed vehicles.

With more than 20 rentals Emmy scooter rentals reached a peak at the launch of the service in July. The number then fell in August significantly and then rose steadily until October. Usage numbers in November and December differ a lot. The assumed explanation for this development is that more vehicles were brought back to the location by the operator in the first weeks after its introduction. Another reason for lower user numbers in November and December is likely to be the colder and thus more uncomfortable weather conditions.

The qualitative evaluation from the residents and business owners and employees around the square was performed via questionnaires from August until the end of December. Paper questionnaires were provided on site and in nearby cafés and shops. In addition, the questions could be answered in the digital version of the questionnaire that was provided on the project website. A total of 33 people replied either on the paper or digital version of the questionnaire, 13 male, 15 female, one diverse and four without indications on gender. Out of these, were 14 residents and 3 business people. With over 55years, the average age was quite high. More than half of the respondents had a higher education entrance qualification (13 people) or high school diploma (nine people). When asked how they asses the new overall situation compared to the situation before, the participants in the survey answered on a scale from 1 to 5 - Much better (1), A little better (2),

Same (3), Slightly worse (4), Much worse (5) with a clear majority "Much worse" (24 of the respondents). In very few cases, however, this was related to the specific mobility services. Interestingly, only four of all respondents used the new mobility services, three people in their free time and one person to go to work. In particular, the position of the mobility offers met with resistance: 23 people stated that the new mobility services would not fit into the local surrounding. When asked for reasons, people indicated that the new service is an impediment to the market (8x); the container is ugly (3x), there is no need for the new services (7x), it leads to a loss of car parking places (7 x). In addition, almost half of the participants in the survey (16 people) stated that they would also not like to have any of the mobility offers at other nearby locations.

Participant observation was performed through regular visits on site by the evaluation team, on different days of the week and at different times of the day. Observation focused on the new parking situation from the start which did not change a lot during the first weeks after the start in June. Most car drivers ignored the newly introduced parking limits and designated spaces for alternative mobility services. In consequence, larger signage in accordance with the Road Traffic Act, had been installed. After a couple of days, the new parking arrangements had been largely accepted by users. But the number of shared cars available on site varied considerably, at times there were no cars available at all. After consulting the mobility providers, the availability of car sharing could be organized more evenly.

The number of available shared bikes, e-mopeds and e-scooters also varied significantly over time. The latter usually outnumbered any other form of mobility - sometimes early in the day, some 30 vehicles were available on site - which was a very large number for this rather peripheral location of the Berlin city center.

A weekly market – existing since many years - has been held right next to the experimental area. On the two market days, access to the mobility services was also massively restricted by market vehicles and visitors. As a result, the access to the vehicles on the market days was restricted. Among other things, this meant that the usage figures on these days were significantly lower than on other days of the week.

4   Results - Process Evaluation

As a follow-up to discussions that took place between residents and organizers at the opening event of Mieri-Mobil, the first user participation activity took place in form of a first residents' workshop. It was organized in August 2021 with the same people, some other interested residents – altogether almost 15 - the district councilor and four project members. During the discussion, four main points emerged, that were subject of complaints: The container that had been set up to serve a storeroom and as a surface for attaching project posters, the massive number of e-scooters, the lacking possibility to participate in the first planning process of the project, and the density to the weekly market.

Regarding the container, it was suggested by all residents that it should be completely removed from this very place for aesthetic reasons. Also, almost all agreed that the e-scooters were considered a problem as the (sometimes) large number of vehicles often looked messy, the users did not follow the rules and the (nightly) battery replacement was a source of noise that has been perceived as very annoying. It was also noted that the project as a whole should have been better communicated. Also, some of the participants expressed that they could not understand the meaning of the new signs that had been put up. Some of these signs have been just recently introduced with the reform of the official national German road traffic regulations ('Straßenverkehrsordnung'). Especially the new sign for car-sharing was not being understood as such. Also it seemed to be hardly comprehensible how the signs for regulating parking spaces exactly were meant and what is the purpose of the container. In this context, it was also criticized that there was no opportunity to discuss the type and scope of the individual measures. The new mobility services were perceived more as a hindrance to the weekly market (Wednesday and Saturday mornings), since they made it more difficult to deliver the goods and also to use the market "as a meeting place". It was therefore proposed to relocate the car-sharing spaces.

The second resident workshop was designed as a simulation game on the future of local mobility. It took place in September 2021 and acted as a follow-up to the first workshop with almost the same group of participants (12 residents, district councilor, three members of the project team, and two external facilitators). In contrast to the first workshop, the participants now agreed that alternative mobility offers make sense - only the implementation at this specific location was criticized partially: These were in particular the insufficient number of cars available and the use and appearance of the container). A relocation of the mobility station to a less exposed part (not on a main through road) was suggested. Also, some participants preferred a distribution of several mobility stations throughout the wider area. It was mentioned that commercially used areas (for example supermarket parking lots) appear to be particularly appropriate locations for new mobility services, since there, they cause less disadvantages for neighboring residents.

A Process Evaluation Workshop was arranged in October 2021 with the four members of the core project team. As all of them have been involved since the beginning they were perfectly suited to comment on the processes of planning and implementing the experiment project. The workshop started with conjointly setting up a timeline of important project-related milestones. Subsequently, these were assigned to events that had either a positive or negative effect on the project:

During the first months of preparation (January to March 2021), the plan to use the entire stretch of road as an experimental space and to use it for activities throughout the project period seemed very ambitious to the project team, especially given the short preparation time and limited resources. However, regular team meetings with members of the project team and responsible persons in the district were able to allay many concerns. In addition, flexible adjustments to the planning and implementation concept could be discussed and decided at the meetings. Holding these regular meetings was considered essential.

In April and May, the lack of project resources became noticeable: Additional costs for necessary traffic signs, for additional time required for informal discussions with residents and the market management or the organisation of further events could not be foreseen prior to the start of the project, and additional funds could not be provided. In consequence, some necessary steps had to be taken without funding but with flexible approaches of the project team. Furthermore, the decision was taken by the local administration that a temporary closure of the street was not feasible. This went along with another decision that there would be no play street because the market could not be relocated and construction site nearby that has not been communicated early enough aggravated the local parking situation. The offer for residents to use a replacement parking space about 200 m away was not accepted. On the one hand, the distance was judged to be too impractical, on the other hand, the insufficient lighting in the replacement parking lot was rejected due to safety concerns, especially from women.

From the very beginning, a sceptical attitude towards the project and the team was expressed by local residents and traders. The most common concerns were fears about restrictions on the usability of the traffic areas and a devaluation of the appearance of the square due to the large number of mobility services. The project team reacted to this scepticism with more intensive communication about the scope and goals of the project. This was done through informal on-site discussions, in written form on a provided poster and the project website, and most importantly in the workshops with local residents. Around July, a positive change in the attitude of many local residents could be observed. Three factors that caused this change need to be mentioned: First, the presence of the responsible district councillor at all workshops, second a good preparation and moderation of the workshops that made the participants feel that their perspectives and suggestions were being heard. Third, the project team invested more time than planned in building trust between the team, the local administration and the residents through informal and formal discussions.

The final step of the process evaluation activities undertaken by the project team was to present the results and findings from the process and impact evaluations to experts from similarly experimental projects in Berlin. In a Peer Review Workshop carried out in February 2022, crucial aspects of implementing mobility experiments were discussed in a group of seven experts and project members. Referring to the intermediary findings from the Mieri-Mobil project three aspects were identified beforehand: Communication (before the start and in the project), organizational pitfalls when implementing experimental spaces, and appropriate evaluation methods.

Regarding the communication aspect, the participants agreed that any project aiming at the transformation towards more sustainable urban mobility must expect strong headwinds. In order to deal with this adequately, transparent communication is necessary right from the start. However, the participants also agreed that very early information concerning the idea to implement a project such as this, led to strong protests, while information about four to six weeks in advance led to productive discussions. This is the case particularly when discussion take place in smaller rounds. This raises the question again - which remains unanswered even after all the decades of successful citizen participation - when is the ideal time to start successful communication.

Furthermore, it is essential that an experimental area is communicated as a "temporary" area and that the term "Temporary" needs to be taken seriously by the organizers. This includes clearly communicating the end of the project right from the beginning. In addition, it is essential to address all local stakeholder groups, i.e. not only residents but also tradespeople and organizations. Personal contacts and discussions seem preferable.

In addition, it could be observed that the number of complaints seems to be higher when people get the chance to give their feedback online.

Challenges to implementing such experiments were distributed responsibilities in the district administration. In the case of Mieri-Mobil, the people responsible for authorizing the restructuring of the parking spaces were different from those setting up the construction site, or in charge of the local market. In addition, many resources are necessary for the preparation and implementation of such projects. This is often exacerbated by time pressure. If Mieri-Mobil had been implemented as originally planned, even more resources would have been needed. As in many other similar projects, the budget was restricted which did not allow for integrating local NGOs with experience in interactive project communications.

Another challenge was that the existence and implementation of the legal basis the 'experimentation clause' were unknown in large parts of the administration.

In addition, the signage with the car-sharing parking space symbols that were new at the start of the project was unknown. This meant that in the first few weeks it was very time-consuming and a lot had to be explained.

A last but important finding was that the degree of car use and car ownership in the respective districts seems to have a surprisingly high influence on the positive or negative effects of mobility experiments. It is also easier to persuade people with higher education (university degree) and (high) environmental awareness to participate.

As mentioned before, the evaluation of such experiments is not common yet – neither the evaluation of the project nor the evaluation of individual measures. Accordingly, there were only few ideas which methods would be suitable. However, it became apparent that older people tend to take part in surveys and reflection. This raises the question of how different user groups could be included in an evaluation.

5   Challenges to Making Digital Mobility Accessible and Inclusive

The last chapters described main results of both impact and process evaluation of the Mieri-Mobil project. Most obviously, during the short running time of the experiment, many adjustments to aims and processes of the project had to be made. Nonetheless, the project can be considered a success, as valuable insight into the usefulness and spatial impact of a mobility station could be gained that will be helpful to organize and implement future mobility experiments in Berlin or elsewhere.

At the same time, it became clear that there are still several challenges to mainstreaming alternative forms of mobility.

From the data available, three basic findings could be identified: First of all, the overall usage numbers increased over the course of the project, with a peak shortly after the introduction of the mobility offers in July. From this it can be concluded that people generally seem to be curious about new mobility offers – interestingly, e-scooters seemed more popular than e-mopeds and shared cars. In order to be accepted such new offers and experiments need time to unfold positive effects.

Second, the usage numbers of open vehicles such as e-scooters or e-mopeds seem to be influenced significantly by external influences such as weather, holiday seasons and pandemic events. In consequence, usage numbers were lower during periods of bad weather, during public and school holidays and during phases with increased Corona rates in November and especially in December.

Despite the fact that most mobility alternatives offered at Mierendorffplatz are considered "smart" ones, it is astonishing how little scientific knowledge has been gained so far about their actual usefulness. Of course, data protection is very important, but it would be very helpful if forms of cooperation between mobility providers and scientists could be developed that would allow for further insight into the actual usability of such mobility offers.

During the evaluation, it was very difficult to obtain reliable data regarding usage numbers. Not all providers were able - or willing - to provide usage numbers, not even after repeated requests. In addition, where data was available, the periods of collection were different. Statements about the usage figures over the entire campaign period were therefore not possible. Likewise, no statements could be made about which persons used the offers, since this information was not transmitted for data protection reasons. In consequence, it is impossible to tell which population groups used the offer and which didn't – and why.

Public attitude towards the project changed over time towards the positive. Most significantly, a bias between users of the mobility alternatives and the non-users could be observed. The latter, the non-users, were basically immediate residents and business people - middle-aged and well-educated - who never intended to use alternative forms of mobility. In consequence, the mobility offers were met with a lot of scepticism regarding their usefulness and – more important - their negative impact on urban space and community structure in the beginning. It may be concluded that addressing questions of adequate access to alternative mobility an of impact of mobility stations on public space - are equally important for the success of such experiments. In addition, in may be concluded that the introduction of a mobility station is much more than just offering alternative forms of mobility and that communication with (potential and actual) users and non-users must be organized very differently: The acceptance of the non-user/residents depended heavily on the degree to which they felt understood and to the extent that suggestions for change and improvement were integrated into the project.

Despite all efforts, not all uses of urban space could be integrated into the project adequately: Above all, this was the conflict between the market on Wednesday and Saturday mornings and the mobility offers. In addition, the positioning of the mobility offers and parking spaces right next to a busy main thoroughfare created some issues with people's safety, be it users of the offers or people walking by. In addition, a road construction site that the project team was not aware of in the beginning, caused further issues with public safety and parking space.

In the future, it will be crucial for any mobility experiment to take into account possibly conflicting uses of urban space or temporary restrictions. This requires an early exchange with different administrative units and local stakeholders. In addition, sufficient preparation time for both implementation and evaluation is crucial, along with a comparatively high input of human resources and dedication of the project team as well as the local administration in order to deal adequately with the experimental character. This must include regular project meetings to coordinate feasibilities and pitfalls, as well as the readiness of the time for flexibility, spontaneous actions, alternative solutions and meaningful communication with stakeholders, businesspeople and residents.

Decadelong experiences with participatory processes emphasized the need for early participation and communication in planning and implementation processes (Creighton 2005). During our peer-review workshop, it became obvious that information, participation and communication in temporary experiments must be framed differently: As experiments are clearly time-restricted and have a test character, it must be clear to everyone that there are not and that there cannot be any prefabricated solutions. And that experiments are not (yet) part of traditional planning and implementation processes. In consequence, communication in experimental projects can start too early. This especially if early phases of communication are not attended to intensively by the project team: In the case of mobility, which is a very emotional and contested topic in Germany, informing early about mobility experiments often triggers negative emotions and may also leave a lot of time for organizing protests. About six to eight weeks before the implementation seemed a suitable period for several representatives of Berlin mobility experiments. In this context, the two workshops with residents can be considered a success as continued dialogue, individual information and commitment of the project team and the district councillor helped to develop informed decisions among the residents and to obtain a different perspective on temporary experiments: "First, I was really annoyed, now I see that something is happening. I think that's good!".

6 Conclusions

It is more than obvious that we, as a society, will need new approaches to urban mobility. The project and its evaluation give concrete impetus as to where and how mobility measures can successfully be organised and implemented. In addition, there were direct consequences of the project and the workshops: The offered range of new mobility services will be relocated to a safer location nearby. As a side effect, other local mobility issues were also discussed in the workshops, for example that a pedestrian crossing is needed nearby and that the general topic of mobility has been picked up by local initiatives.

Mieri-Mobil presented a specific mobility mix at a specific location. Experiences made with each experiment are unique as they relate to specific administrative settings, to a specific population and stakeholder structure, and to specific urban space. However, some of the results can be transferred to other settings in Charlottenburg-Wilmersdorf or to other sites in Berlin (see Sect. 5).

Therefore, for future similar experiments in the district, cross-departmental reflection on the project would be recommendable. A comprehensive (sustainable and) smart urban mobility system will need more integration (Uteng et al. 2019) and coordination (Docherty et al. 2018) in order to overcome fragmented responsibilities, singular solutions and path dependencies. In addition, a continuous exchange with those involved in other experimental spaces in Berlin and elsewhere seems sensible, because similar experiences to Mieri-Mobil might have been also made in other projects.

It remains unclear in what terms such experiments focussing on alternative, smart mobility is accessible to different population groups and therefore (socially) inclusive. Clearly, using the mobility alternatives costs money that not everyone is willing or able to pay. In addition, the concentration of mobility services in one single spot leaves many people with long (walking) distances – which may discourage people to use these offers, especially if they have problems walking. And there are indications that people with lower education are less interested in questions of mobility (Neue Mobilität Berlin 2020). Mobility alternatives in the project were clearly supply-led and focussed on testing new technologies. In consequence, mobility needs of specific user groups may not have been met.

Nonetheless, Mieri-Mobil clearly showed the importance of experimenting as the organisation of uses of urban space is very complex, and in most cases unintended effects and overlooked uses of space may complicate a proper integration of new forms of mobility into people's daily lives.

The project also showed that a much more and more in depth evaluation of these projects are needed to be able to really address not only the mobility needs of specific user groups but also to guide planners and administrations the direction towards reaching a true mobility transformation.

References

Abraham, M., Rösler, M., Kreutz-Hassinen, E.: Cities.multimodal Evaluation Report (2021). https://www.cities-multimodal.eu/sites/cmm/files/materials/files/cmm_evaluation_report_2021.confirmed.pdf. Accessed 17 Nov 2021

Arseneault, D.: Mobility Hubs: Lessons Learned from Early Adopters (2022)

BA/Bezirksamt Charlottenburg-Wilmersdorf von Berlin 2017: Grobcheck Stadtumbau Mierendorff-INSEL Charlottenburg-Wilmersdorf, Berlin (2017). https://www.stadtentwicklung.berlin.de/staedtebau/foerderprogramme/nachhaltige-erneuerung/fileadmin/user_upload/Dokumentation/Projektdokumentation/Charlottenburger_Norden/PDF/2017-Mierendorff-insel-Grobcheck-Stadtumbau_01.pdf. Accessed 12 Feb 2020

Bell, D.: Intermodal mobility hubs and user needs. Soc. Sci. **8**(2), 65 (2019)

Creighton, J.L.: The Public Participation Handbook: Making Better Decisions Through Citizen Involvement. Wiley (2005)

Docherty, I., Marsden, G., Anable, J.: The governance of smart mobility. Transp. Res. Part A: Policy Pract. **115**, 114–125 (2018)

Difu (Deutsches Institut für Urbanistik): The Multimodal Future of On-Street Parking. A Strategic Approach to Curbside Management. Difu, Berlin (2020)

Hadachi, A., Lind, A., Lomps, J., Piksarv, P.: From mobility analysis to mobility hubs discovery: a concept based on using CDR data of the mobile networks. In: 2018 10th International Congress on Ultra Modern Telecommunications and Control Systems and Workshops (ICUMT), pp. 1–6. IEEE (2018)

NMB Neue Mobilität Berlin: Mobilität ohne privates Auto erleben. Ergebnisse der Berliner SOMMERFLOTTE. In: Stein, T., Bauer, U. (Hrsg.): Bürgerinnen und Bürger an der Verkehrswende beteiligen. Erkenntnisse, Erfahrungen und Diskussionsstand des Städtenetzwerktreffens aus dem laufenden BMU-Forschungsprojekt City2Share und kommunaler Umsetzungspraxis. 3. City2Share-Diskussionspapier, Berlin, pp. 30–33 (2020)

Pangbourne, K., Stead, D., Mladenovic, M., Milakis, D.: The case of mobility as a service: a critical reflection on challenges for urban transport and mobility governance. In: Marsden, G., Reardon, L. (eds.) Governance of the Smart Mobility Transition, pp. 33–48. Emerald Publishing, Bingley (2018). ISBN 978-1-78754-320-1

Rupprecht Consult (ed.): Guidelines for developing and implementing a sustainable urban Mobility Plan, 2nd edn. (2019). https://www.eltis.org/sites/default/files/sump-guidelines-2019_mediumres.pdf. Accessed 07 Apr 2020

Schelling, J.: Mobility hubs: how will they function, look and enrich the city (2021)

Uteng, T.P., Singh, Y.J., Hagen, O.H.: Social sustainability and transport: making 'smart mobility' socially sustainable. In: Urban Social Sustainability, pp. 59–77. Routledge (2019)

Utriainen, R., Pöllänen, M.: Review on mobility as a service in scientific publications. Res. Transp. Bus. Manag. **27**, 15–23 (2018)

van Gerrevink, I.: Ex-post evaluation of neighbourhood mobility shared mobility hubs: a qualitative research on the factors influencing the usage and effects of mobility hubs (2021)

User-Centric Design for Digital Mobility

Universal Design and Transport Innovations: A Discussion of New Mobility Solutions Through a Universal Design Lens

Jørgen Aarhaug[✉]

Mobility, Institute of Transport Economics, Oslo, Norway
Jorgen.Aarhaug@toi.no

Abstract. Most technological advances in mobility result in better accessibility for many, yet the benefits remain unevenly distributed. Universal design is a strategy to counter social exclusion, involving the design of products and environments to be usable by all people, to the greatest extent possible, without the need for adaptation or specialised design New and improved mobility technologies typically result in increased mobility. However, most new technologies create both winners and losers – and who wins and who loses depends on how the mobility solution in question is introduced to the mobility system. This study finds that many of the new mobility technologies that are introduced, though not directly relating to universal design, strongly affect the universality of access to mobility. The chapter aims to give insight into how certain new mobility solutions affect different user groups, and to highlight how the outcome is a function of the interplay between technology and its implementation. The paper concludes by pointing at the need for regulation to align the objectives of the actors behind new technologies and an inclusive society.

1 Introduction

Technological change is an important driver of increased welfare. As society becomes ever more interconnected, mobility is an increasingly important precondition for functioning fully as a citizen. This chapter looks at the relation between mobility innovations – specifically, innovations facilitated by digitalisation – and universal design (UD).

The concept of UD in reference to a strategy towards promoting social inclusion was first coined by the architect Ronald Mace, who defined it as *'the design of products and environments to be usable by all people, to the greatest extent possible, without the need for adaptation or specialized design'* (Mace 1998). The term is used primarily in the United States, Scandinavia and Japan, while the expression 'design for all' is used with a similar meaning elsewhere (Audirac 2008). The term is also used in the UN Convention on the Rights of Persons with Disabilities, in which it is defined as *'the design of products, environments, programs and services to be usable by all people,*

This chapter was written in parallel with Aarhaug (2022) a Norwegian-language article, for the edited volume "Universell utforming i transportsektoren" (Fearnley and Øksenholt 2022).

© The Author(s) 2023
I. Keseru and A. Randhahn (Eds.): *Towards User-Centric Transport in Europe 3*, LNMOB, pp. 157–172, 2023.
https://doi.org/10.1007/978-3-031-26155-8_10

to the greatest extent possible, without the need for adaptation or specialized design. "Universal design" shall not exclude assistive devices for particular groups of persons with disabilities where this is needed' (United Nations 2010).

The UD concept has been adopted for use in transport, internet and communication technologies (ICT) and education following its introduction into the built environment. Within architecture, similar concepts can draw on history back to the 1970s. According to Story et al. (1998), early efforts to render environments accessible frequently depended on segregated measures that were *'more expensive and usually ugly'* than UD, which includes accessibility for all in early design phases.

Within mobility, the UD concept has mainly been used in relation to public transport (Audirac 2008). Bjerkan (2022) finds that half of the documents, journal articles and reports relating to barriers and transport are based on public transport cases.

Since 2000, there has been robust research interest in how mobility restrictions can be a cause of social exclusion (Cass et al. 2005; Preston and Rajé 2007; Preston 2009). These studies argue that mobility constitutes a barrier that prevents individuals from fully participating in the normal activities of society, despite their desire to do so. This conceptual work on social exclusion has increasingly been joined by empirical studies focusing on identifying individual barriers, and how these can be mitigated – as illustrated by a recent literature review by Bjerkan (2022).

The concept of UD, when used in the context of transport, is a way of thinking about these issues mainly as an alternative and complement to 'accessibility'. Here, the difference can be interpreted as accessibility with a focus on solutions created for individuals with impairments, while UD is focused on providing a solution in which impairments are irrelevant: in other words, where the solution can be used by as many people as possible, impairment or no.

The UD philosophy is somewhat in contrast to the commercial and technological focus of new transport innovations. This paper aims to explore this potential conflict, using certain innovations in Norway in 2020 as a case.

1.1   A Window of Opportunity for New Mobility Solutions

Society is changing. In the broader sense, we are looking at a reorganisation of goods and services available following the fifth technological revolution (Perez 2003), of which digitalisation is a core component. This also affects mobility. In addition, there is an established understanding that the centrepiece of 20th-century mobility – the private car – represents an unsustainable way of providing mobility (Sheller and Urry 2006; Geels et al. 2012). This combination of digitalisation, which increases the opportunity space for innovation, and the narrative placing the private car as the centrepiece of our day-to-day mobility is challenged, creating a window of opportunity for new mobility solutions. Many have emerged to fill this space.

Looking back, the private car powered by an internal combustion engine was one of the most important technological advances of the 20th century, and the most important change within day-to-day mobility. The car solved the problem of horse manure in city streets, and the technology provided an enormous growth in individual mobility. On the flip side, it also brought new challenges related to traffic safety, noise, urban sprawl and consumption of fossil fuels – and conditions related to inclusion. For a variety of reasons, including age, health, ability, wealth and ideology – large segments of the population do not have access to their own car. The car-centred mobility system, is therefore not neutral.

How people relate to cars depends on a series of factors including life events (Uteng et al. 2019a) and wealth (Bastian et al. 2016). The position of the car as the centrepiece of the mobility system is thus being challenged along a series of dimensions, both of which are related to environmental concerns (Geels et al. 2012) and to the idea of what constitutes a good life (Schwanen et al. 2015). In this context, UD and public mobility solutions become more important.

This paper draws on the technology mapping conducted by the Norwegian Board of Technology in 2020 (Haarstad et al. 2020), and experience from empirical studies of how people with mobility impairments interact with new technology – in particular, (Øksenholt and Aarhaug 2018). The mapping exercise was conducted to chart which innovations were likely to influence mobility in the context of Norwegian cities. The technologies listed include physical solutions, such as e-scooters and autonomous vehicles, but also new ways of offering mobility, such as mobility as a service (MaaS) and cooperative intelligent transport systems (C-ITS). Although mapped and analysed in a specific context, the technologies are common in many parts of the world. A selection of these technologies is explored further in this paper. The empirical experience used to discuss these technologies is mainly drawn from research projects in which the author has had a role; these projects were centred on individual issues. This experience is supported by prior research, including (Bezyak et al. 2017, Deka et al. 2016, Fearnley et al. 2022.b, Nielsen 2021).

2   New Mobility Solutions

There is no single definition of what new mobility solutions are. For the purposes of this paper, there is no need for a precise definition, as the list of technologies discussed is not exhaustive. It focuses on technologies that are either established 'on the streets' following 2010 or expected to have a noticeable presence in northern European cities before 2030, drawing on the findings of expert panels and reports (Haarstad et al. 2020, Kristensen et al. 2018, Bakken et al. 2017, Aarhaug et al. 2018). These sources include discussion of a wide range of solutions, but there are common features. Digital technologies constitute a substantive element in many of the technologies; these work as facilitating or general-purpose technologies across sectors (Bresnahan and Trajtenberg 1995). In addition, several mobility-related innovations are not 'new' transport services per se, as the physical mobility solutions are often very similar to the pre-existing solutions. Rather, existing mobility options are improved (increasing efficiency) and offered in new ways through different business models, facilitated by the development of digital technologies.

Digitalisation increases opportunity space, as do new broader technologies in general. This is not the same as stating that innovations always make things better: neither technology nor how it is adopted are neutral concepts – new technologies create both winners and losers. Moreover, an individual may be both a winner and a loser, when different measurement criteria are applied. That a technology increases total welfare in society does not mean that the benefit is evenly distributed. Indeed, it often is not. Further, it is not certain that those who most benefit from the new technology will be able (or willing) to compensate those who lose out. As a phenomenon, the use of a new technology is not necessarily distribution- and inclusion-neutral, and how a new technology is introduced to market is not accidental.

2.1 Technology Uptake

Innovation studies have given rise to several conceptual models of how new technology is diffused. One of the classic and most widely used and criticised is that developed by Rogers (2010/1962). Within this model, the uptake of the new technology follows an s-curve, where a technology moves from being a small niche phenomenon – with users often labelled as 'trendsetters', or members of the 'urban elite' – through a take-off and acceleration phase; the technology is then gradually adopted by the rest of the population. This model of adoption contrasts with the philosophy of UD: Rogers' model is an attempt to theorise based on observations, while the concept of UD is inherently normative.

The idea that some forward-leaning, or 'elite', individuals use a mobility solution before it spreads to other parts of the population does not have to be a problem in terms of UD. But if the elite's consumption cannot be replicated across the population, it becomes problematic in terms of UD because it reflects different access to mobility. As an example, a new mobility solution may require a specific type of smartphone (as some hailing services do), payment by credit card (as many private companies require), substantial income (to afford the service) and a driver's license (in the case of car sharing). A new technology that is only useful for a few will thus not necessarily be universal, and may be at odds with UD as a policy objective.

Avoiding this bias can be difficult. Many of the new mobility technologies that have come on the market are initially aimed at typical 'early adopters'. These are individuals who also tend to have demographic characteristics that overlap with those who initiate the new technologies. In addition, many new technologies are developed in and for a world market. Low replication cost is a key component of digitalisation. User participation may be limited to the question of how the technology should be introduced in the specific locality, rather than how the solution is designed. This may reduce the scope for adaption to address the needs of other user groups than those upon which the developers initially focused.

2.2 Examples of New Technologies

A common denominator for the technologies selected in this work is that they have been influenced by the development of digital technologies. In this way, and as mentioned earlier, they can be labelled as part of the fifth technological revolution, following Perez' (2003) classification. Digitalisation can be understood as the way digital technologies are introduced into society. Here, an important component is how information is transferred from being a physical to a digital entity – transformed from atoms to bits (Negroponte et al. 1997). This means that replication costs associated with information drop dramatically, and the movement of information is decoupled from the movement of physical entities. This allows a series of new services, and new ways of offering pre-existing services. Information, including vehicle location and status, can be made available for existing and potential travellers at low cost, reducing the disutility of travelling (Flügel et al. 2020).

Digitisation thus provides an opportunity to establish new service offerings based on available information both commercially and non-commercially. This helps potential travellers make more informed choices about when, where and how they travel. At the same time, it can also help to increase the divide between those who have access to this information, for example through the use of smartphones, and those who do not. This points back at the concept of universality and main solutions. Does it mean that the mobility system should include mobility solutions for all, or that each element of the mobility system should be accessible for all?

Predicting the future is challenging. In relation to mobility, the practice has long been to make predictions based on modelling along established trends. This approach has been criticised for creating lock-ins in established technologies. To address how new technologies play-in in future mobility, several methods have been used, including modelling with (very) alternative assumptions, backcasting and scenario building. For the purpose of creating a coherent discussion, this chapter makes no independent effort to assess which technologies are relevant. Instead, as noted above, it uses a list identified by the Norwegian Board of Technology as a starting point (Table 1) and adds to this by using examples and assessing UD relevance. In Table 1, UD relevance is judged based on discussions between the author and other researchers with experience from UD and technology implementation.

Table 1. Transport innovations adapted from (Aarhaug 2022)

Technology	Status (2022, Norway)	Examples	UD relevance
Digital transport systems			
Mobility platforms/Mobility as a Service (MaaS)	Pilot/upscaling	Whim, Bolt, various apps and projects from PTA[a]s	Large
Cooperative intelligent transport systems (C-ITS)	Different stages		Large
Micromobility			
Electric bikes and e-scooters	Established	In common use	Some
Shared micromobility	Established	VOI, Urban Sharing, TIER, BOLT etc	Some, most discussion from externalities (misuse)
Autonomous micromobility	Experimental		Potentially large

(*continued*)

Table 1. (*continued*)

Technology	Status (2022, Norway)	Examples	UD relevance
Car/taxi			
Electric vehicles (EVs)	Established	Battery electric vehicles (BEVs) from most producers	Some
Car sharing	Established	Bilkollektivet, Hertz-bilpool, Hyre etc	Some
Taxi apps (ridesourcing, ridehailing, TNC[b]s etc.)	Established	Uber, Bolt, Yango, MyTaxi	Some
Ridesharing	Established	GoMore, BlaBlaCar, various	Small
Autonomous vehicles	Pilot	Waymo	Potentially enormous
Taxi drones	Pilot	EHang	Small
Public transport			
On-demand bus services (DRT)	Established	Various	Large
Autonomous small buses	Established		Large
Autonomous bus fleets	Pilot		Small
Autonomous ferries	Pilot		Small

[a]PTA - Public transport authorities.
[b]TNC - Transport network companies.

In Table 1, small UD relevance means that the technology is judged to not directly impact UD, and thus is less relevant for UD policies. Some UD relevance means that the technology impacts mobility in a heterogeneous way (mainly by providing advantages to some users and possible disadvantages to others), and that this differentiation is linked to users' characteristics. The differentiation may not be directly related to mobility impairments, but is changing the mobility market in a way that influences persons with disabilities. An example would be a reorganisation of the taxi/non-emergency vehicle-for-hire markets, by removing requirements for operators to provide wheelchair-accessible vehicles. Large impact is when the technology is judged to influence persons with mobility impairments directly.

The following text focuses on the technologies that are expected to be most relevant in a UD context.

Mobility as a service (MaaS) is a digital platform that connects various mobility offerings from different modes through a single user interface. In this innovation, the main issues are related to implementation, not the development of the technology. As pointed out by Smith and Hensher (2020), MaaS actors have had more success in developing the technology than in functioning as economic and organisational entities. Theoretically, MaaS should increase the possible user group for a particular mobility mode, through reducing the barrier created by lack of information and creating a possibility for nudging; it is possible to inform travellers of various characteristics of the service in question at lower cost. The drawback, in a UD context, is that MaaS requires digital skills and smartphone access. Other potential issues are related to a fragmentation of responsibility: this issue arises when the operator providing the service is not the same as the one interacting with the customer at the point of booking.

Cooperative intelligent transport systems (C-ITS) refers to transport systems where two or more sub-systems are able to communicate. This may include vehicles that can communicate with other vehicles and/or infrastructure components. This is not a single technology, rather it is a set of technologies that can gradually contribute to more interconnected mobility systems and automation. C-ITS can help to make mobility more universally designed, by providing access to more and better information about real-world events in the system. An example of this is geofencing, which can limit access to dynamically defined zones: regulating speed and enforcing parking restrictions for electric scooters, introducing zero-emission zones etc. Another example of C-ITS are 'beacons' that can make time- and place-specific information about the mobility service available for visually impaired people.

Micromobility is a common term for small vehicles, including e-bikes, e-scooters and skateboards (Fearnley 2021). Some are designed to be used in mixed traffic with pedestrians. To the extent that these vehicles replace cars and vans, they can contribute to make the street space more available to softer travellers. However, when introduced to pedestrian areas, they typically increase the weight and speed of vehicles in these areas.

E-bikes make biking more accessible, and enable a wider segment of the population to bike further (Fyhri and Sundfør 2020). E-scooters provide access to individual motorised mobility for persons who would otherwise have less access to motorised mobility, being cheaper than taxis and private cars and more available than public transport. For persons with disabilities, issues with e-scooters are largely related to parking. That these small vehicles are left on the pavement is a problem, as they may get in the way of wheelchair users and can be a danger to the visually impaired.

Shared micromobility consists of bicycles, e-bikes or e-scooters that can be rented via subscriptions or on a per-trip basis. This decouples ownership and use and is expected to improve access and reduce the threshold for using the technology. Still, user surveys indicate that the majority of the users are young, wealthy, without disabilities and using the services in city centres (Fearnley et al. 2022a).

Autonomous micromobility represent a future iteration of small vehicles. It is still in the concept phase but has the potential to solve many of today's issues with micromobility. Having the vehicles drive autonomously may facilitate access to the service, including for the visually impaired. It may also potentially reduce the issues with misplaced bikes and e-scooters.

Electrification helps to make cars less polluting. By itself electrification has little effect on UD and accessibility. Still, battery electric vehicles (BEV) can serve as an illustration of how new technology is introduced to the market, without taking UD into account. The first BEVs that came on the market were only suitable to meet the needs of a small segment of the population. There were few models, with a short range, high purchase cost, and limited publicly available charging points. As the technology has become more mature, more models are available and BEV can cover a wider range of needs. Although they can replace internal combustion engine vehicles (ICE), BEVs are still cars. Charging – especially rapid charging – requires a relatively functional person to operate the charger. Moreover, driving requires a license, and the cost of owning and operating a vehicle exclude many.

Car sharing enables car access without having to own a car. In practical terms, car sharing reduces the barrier for each trip, compared to car renting, while still having a higher barrier than private car ownership. Car sharing can reduce car ownership, parking needs and emissions from car ownership in urban areas (Chen and Kockelman 2016). This can help free up space for other types of road users and have a positive effect on accessibility. At the same time, it is not clear how car sharing will affect city space and car ownership in the long run, since the usage patterns and motivations for participation are still under development (Julsrud and Farstad 2020). The implications of car sharing for UD are also uncertain. Car sharing is aimed at people who are able to drive cars with a standardised design and exclude people who cannot use such cars. In this way, it can be argued that car sharing may increase the differences between those who are 'inside' and 'outside' the norm. However, this line of argument seems a bit extreme.

Ridesourcing is one of many terms used to describe new, platform-organised, taxi-like services. Other terms include transport network companies (TNCs) and ridehailing. The main effect of these services has been to make taxi travel – traditionally the most accessible form of motorised mobility – available for more people. The services are also generally perceived as safer than pre-existing services, further reducing the barriers to use (Aarhaug and Olsen 2018). The potential downside is linked to reduced scope for local authorities to regulate the supply, which may (as is the case in Norway) reduce the number of wheelchair-accessible vehicles (Aarhaug et al. 2020). This raises the question concerning at what level a system should be accessible. Is it sufficient to have access to some vehicles, or does every vehicle in a fleet need to be accessible? The latter would be more expensive and likely require some form of economic transfer, as the market solutions seem to focus on a narrower user segment than what UD dictates.

Autonomous cars have the potential to radically change the mobility system (Docherty et al. 2018, Nenseth et al. 2019). An expectation is that autonomous cars will make car-based mobility accessible to a larger part of the population. In extension, this will lead to an increase in mobility, especially for those who currently do not have access to their own car. Here, downsides include increased traffic and energy use, unless strict policies are introduced. The outcome will be highly policy dependent. Autonomous vehicles may well blur the distinction between private and public transport (Enoch 2015, Seehus et al. 2018). Automation may reduce the cost associated with providing the service, allowing public transport with higher frequency and or more flexibility for similar cost. This should increase the attractiveness of public transport relative to other modes. However, many questions relating to how autonomous vehicles will be perceived and regulated is still unanswered.

Demand responsive transport can be seen as closely resembling MaaS, by making a public transport service available on demand, through a potentially multimodal platform. This should point towards increased accessibility and improved UD. The potential downside is linked to the difficulty associated with communicating such services to vulnerable groups (Skartland and Skollerud 2016). In parallel to other services that rely on the automated processing of bookings, issues may arise concerning a lack of the correct tools or knowledge to order the services.

2.3 Impacts of New Technology

Common across these new technologies is that the innovations are mainly about combining existing elements and services in a new way. Here, the possibility of a better user interface through connection to smartphones has been particularly important. Looking ahead, it seems that autonomous vehicles, in addition to emissions-reducing technology, will also become increasingly important. If autonomous vehicles are used to a greater extent, it will have major consequences – both for how people think about transport and accessibility to mobility services. This is a field that is being researched, but where there is still a great deal of uncertainty.

Lenz (2020) points out that in addition to the obvious gains involving better information flow and greater access to information about transport services, there are many factors related to new transport technology and smart mobility that are poorly elucidated. For example, data flow across systems presents new challenges in terms of risk, ownership and responsibility. Many aspects of new mobility technology influence different users in different ways, potentially creating new inequalities. This applies along several dimensions, and is often under-communicated. The typical user of new mobility services described by Lenz (2020) has many similarities with the typical early adapter in traditional technological transition frameworks: young, wealthy, technology-oriented and able-bodied. Depending on whether and how quickly uptake of the new technology spreads to the rest of the population, this may mean that the segment of the population that has access to mobility services becomes both larger and smaller. The optimistic expectation is that more mobility may be available to more people; the negative expectation is that the differences between people's access to mobility increase, as a result of some gaining access to better services while others retain their current mobility – or lose some of this mobility as users who can select the new services. Still, there are a number of examples of user participation in the development and implementation of new technology in the transport sector, especially related to public transport. There are also a number of technological developments that support inclusion. Examples of such technologies include navigation solutions for the visually impaired on smartphones, and contactless payment using mobile phones, which makes it possible to avoid vending machines for tickets and various forms of driver assistance. It seems that the consequences of new technology are mainly determined by how the technology is used and what frameworks and regulations are established, and not just the technology in isolation. While the opportunity space is increasing, the benefits may not necessarily reach everyone.

3 New Possibilities and Challenges

In studies of travel behaviour, an expected finding is that people with disabilities travel less than people without disabilities (Nordbakke and Schwanen 2015; Aarhaug and Gregersen 2016; Gregersen and Flotve 2021). Some of this is likely explained by correlated variables, such as lower work participation and age, but a major component includes real and perceived barriers (Lodden 2001; Nordbakke and Hansson 2009; Bjerkan et al. 2011; Aarhaug and Elvebakk 2015).

A universally designed mobility system is not a system in which small changes are made to accommodate some individuals with special needs. Rather, UD includes individual design elements that have been tailored to incorporate better accessibility, such as stepless access to public transport, real-time information, automatic or beacon-based onboard information and shelters with seats. These are elements that are highly valued by all users, irrespective of ability (Veisten et al. 2020). There is no contradiction between socio-economically sound investments and UD. However, to some extent, this is challenged by new technology.

There is a correlation between increased income and travel choice. The more affluent are increasingly choosing to travel privately (Button 2010/2014). This need not have a direct impact for UD to the extent that new offers are in addition to the existing ones. But it has the potential to have significant indirect consequences, in that it may undermine the financing models for publicly available mobility solutions, which in turn may be detrimental for universality.

3.1 Concerns Related to Specific Groups

For people with mobility impairments, many mobility-related innovations are good news. Access to electric bicycles can increase the mobility for those who have the opportunity to use them. Public transport services on demand enable door-to-door travel for more people, especially where the first- and last-mile trip segment has been a barrier to travelling. Increased access to car-based mobility also offers increased participation, provided that people with mobility impairments have the opportunity to take advantage of the service – in other words, that the user interface and vehicles are accessible.

For wheelchair users, a transition from bus-based to private-car-based offers can present a challenge, mainly through a reduction in the number of wheelchair-accessible vehicles available. Another challenge is the provision of shared electric scooters and other free-floating mobility systems that can constitute physical barriers in public areas.

For people with orientation impairments, digitisation of information has helped to make travel experiences easier. In practical terms, digitisation of travel information means that the available information – including digital notices and automatic calls – can be (and is) disseminated across platforms and on smartphones, through beacons. Together, this makes the travel experience less intimidating. Digitalisation in a broader sense also makes access to door-to-door transportation more available. While research has identified weaknesses in the implementation of the more advanced information systems (Øksenholt and Aarhaug 2018), this constitutes minor disadvantages of a development that is mainly positive.

For the visually impaired, many of the mobility innovations mean increased access to door-to-door transport. This can be very helpful. The expectation is that this will be even better when autonomous transport is introduced on a larger scale. If autonomous mobility is available, several (but not all) of the mobility barriers that visually impaired people currently face will be eliminated. Digitisation and digitalisation have already helped to reduce barriers for the blind and partially sighted. A challenge for people with impaired vision is that new transport solutions, especially micromobility, mean that traffic in pedestrian areas operate at a higher pace and is carried out with heavier vehicles (electric scooters are heavier than manual scooters, electric bicycles are heavier than non-motorised bicycles etc.). This increases the risk of accidents and the severity of such accidents when they occur. In addition, several of the new transport services are only available through smartphone and app booking. This can be challenging as not all smartphones and apps have sufficient support for text-to-speech programs etc.

A structural challenge in mobility is that increased prosperity generally leads to increased use, and in particular of private transport solutions, such as the private car. This has several effects that affect UD. Directly, this means that the revenue base for the shared solutions goes down, as fewer people pay for tickets. To some extent, this can be mitigated through transfers. Targeted taxes on cars, such as urban toll roads, can create transfers from motorists to public transport users. It can also contribute to providing incentives to not travel by car and to maintaining a better public transport system than ticketing supports. At the same time, this system functions only to a limited extent as a redistribution policy (Fearnley and Aarhaug 2019) and may also have limited effects on behaviour, as it does not mitigate the underlying disadvantage faced by public transport in terms of travel time (Lunke et al. 2021; Lunke et al. 2022).

The challenge with new mobility technologies is that many have significant user cost. This can negatively affect overall mobility in two ways: 1) those who cannot afford it do not have access to the increased mobility that new technology entails; and 2) when parts of the population switch to using new mobility solutions – which are often private and user-financed – it undermines the financing of the mobility solutions on which others depend.

3.2 New Mobility Technology and Universal Design

Several aspects of new mobility technologies are challenging when compared with the policy objective of a universally designed society. How new technology affects the goal of UD is largely related to how new technology is introduced and implemented. On the one hand, new technology increases the opportunity space for actions. On the other, the ability to adopt new technology is unevenly distributed, which may leave many with reduced mobility.

The introduction of new mobility technologies can thus result in new barriers – physical, technological, economic and mental. How this turns out is not only a result of the properties with the relevant innovations, but also the policies surrounding their use. Many of the technological advances that have taken place, especially in public transport (e.g., real-time information, smartphone ticketing and stepless access) have helped to make public transport more accessible. The valuation studies also show that these are measures from which most users benefit, in support of the UD thinking (Veisten et al. 2020; Fearnley et al. 2009).

Looking at technologies that have been introduced under the label of UD in Norway from 2010 to 2020 – in particular, information systems and improved vehicle and travel chain design – these have contributed to creating a more universally designed mobility system (Aarhaug and Elvebakk 2015; Fearnley et al. 2022b). These studies focus on experience within public transport, where important components include real-time information on stops, the use of beacons and improved operator knowledge. This is a great advantage, in terms of a) providing more reliable information to those on board, with automatic stop calls and information screens; and b) through the possibility of creating smart and connected travel planners, and multimodal ticketing systems: measures that are highly valued by all passengers irrespective of impairments (Veisten et al. 2020). However, other components that have been introduced have resulted in new barriers to use. In particular, these are related to the digital skills needed when introducing smartphone or other forms of electronic ticketing (Øksenholt and Aarhaug 2018), but also a wider issue in new mobility (Uteng et al. 2019b).

As argued in this chapter, there are indications that the challenges to UD become more acute with some of the new mobility technologies being introduced. The outcome is increasingly policy dependent.

4   Conclusions

Mobility innovation facilitated by digitalisation has helped to make the mobility system – public transport, in particular – accessible to larger segments of the population. This means moving in the direction of UD. Mobility innovations is helping to facilitate this development: the result is increased mobility and the opportunity for increased community participation.

Several technologies exist as niches that can potentially contribute positively towards UD. Technologies allowing autonomous motorised means of transport in a mixed traffic system have great potential to provide better mobility for all. These technologies can help make mobility currently only accessible by motorists available to more people. This would result in a substantial increase in the opportunity space for people with disabilities. Nevertheless, it could also lead to unfavourable scenarios along other parameters, such as congestion.

Modelling shows that autonomous vehicles can contribute to reduced transport volumes (ITF-OECD 2018), but the conditions for achieving these results are often strict and unrealistic – including the ability to force users away from private mobility and to share, in a way that in day is not feasible within a democratic society. When relaxing these assumptions, the scenarios become far less attractive, from both environmental and societal perspectives (Berge 2019). The question becomes how the opportunities offered by the innovations are used.

Previous research shows that measures for UD have had great socio-economic benefits; however, the same research shows that accessibility is not necessarily something that market players prioritise without being required to do so. This prompts a set of political considerations. The legal framework – at either the national or EU level – can serve as a tool to increase the benefits to all. The actors would be forced to choose the solutions they should choose anyway, if their aim was to maximise the welfare of society.

If the objectives of UD are to be achieved, regulations must be implemented to distribute the benefits from innovations such that those outside the typical early adopter group also benefit from them. This can be achieved by requirements related to the design of new service offerings, such as linking rights to offer a service commercially with an obligation to ensure adequate accessibility, or by imposing taxes and fees on those services that cause inconvenience to others. This revenue can then be used to support the mobility needs of those in society who do not have the opportunity to directly utilize the new technology.

References

Aarhaug, J.: Fremtidige transportløsninger: teknologi, design og innovasjon. In: Fearnley, N., Øksenholt, K.V. (eds.) Universell utforming i transportsektoren. Norsk forening for ergonomi og human factors – NEHF forlag, Oslo (2022)

Aarhaug, J., Elvebakk, B.: The impact of Universally accessible public transport–a before and after study. Transp. Policy **44**, 143–150 (2015). https://doi.org/10.1016/j.tranpol.2015.08.003

Aarhaug, J., Gregersen, F.A.: Vinter, vær og funksjonsnedsettelser – en dybdeanalyse i RVU, TØI-rapport, 1543/2016. Transportøkonomisk institutt, Oslo (2016)

Aarhaug, J., Olsen, S.: Implications of ride-sourcing and self-driving vehicles on the need for regulation in unscheduled passenger transport. Res. Transp. Econ. **69**, 573–582 (2018). https://doi.org/10.1016/j.retrec.2018.07.026

Aarhaug, J., Oppegaard, S.M.N., Gundersen, F., Hartveit, K.J.L., Skollerud, K., Dapi, B.: Drosjer i Norge fram mot 2020, TØI-rapport, 1802/2020, Oslo, Transportøknomisk institutt (2020)

Aarhaug, J., Ørving, T., Buus-Kristensen, N.: Samfunnstrender og ny teknologi - Perspektiver for fremtidens transport, TØI-rapport, 1641/2018, Oslo, Transportøkonomisk institutt (2018). https://www.toi.no/publikasjoner/samfunnstrender-og-ny-teknologi-perspektiver-for-fremtidens-transport-article35026-8.html

Audirac, I.: Accessing transit as universal design. J. Plan. Literat. **23**, 4–16 (2008)

Bakken, T., et al.: Teknologitrender som påvirker transportsektoren. SINTEF (2017)

Bastian, A., Börjesson, M., Eliasson, J.: Explaining "peak car" with economic variables. Transp. Res. Part A: Policy Pract. **88**, 236–250 (2016). https://doi.org/10.1016/j.tra.2016.04.005

Berge, Ø.: The Oslo Study – How Autonomous Cars May Change Transport in Cities, Report, Oslo (2019). COWI I PTV Group https://www.cowi.no/om-cowi/nyheter-og-presse/ny-rapport-slik-vil-selvkjoerende-transport-paavirke-oslo-regionen

Bezyak, J.L., Sabella, S.A., Gattis, R.H.: Public transportation: an investigation of barriers for people with disabilities. J. Disab. Policy Stud. **28**, 52–60 (2017). https://doi.org/10.1177/1044207317702070

Bjerkan, K.: Funksjonskrav for inkluderende transport. In: Fearnley, N., Øksenholt, K. (eds.) Universell utforming in transportsektoren. NHEF – forlag, Oslo (2022)

Bjerkan, K.Y., Nordtømme, M.E., Kummeneje, A.-M.: Transport til arbeid og livet. Transport og arbeidsdeltakelse blant personer med nedsatt funksjonsevne [Transport to life and employment. Transport and labour market participation in people with disabilities]. SINTEF Technology and Society (2011)

Bresnahan, T.F., Trajtenberg, M.: General purpose technologies 'Engines of growth'? J. Econ. **65**, 83–108 (1995)

Button, K.: Transport Economics, 3rd edn. Edward Elgar Publishing (2010/2014)

Cass, N., Shove, E., Urry, J.: Social exclusion, mobility and access. Sociol. Rev. **53**, 539–555 (2005)

Chen, T.D., Kockelman, K.M.: Carsharing's life-cycle impacts on energy use and greenhouse gas emissions. Transp. Res. Part D: Transp. Environ. **47**, 276–284 (2016)

Deka, D., Feeley, C., Lubin, A.: Travel patterns, needs, and barriers of adults with autism spectrum disorder: report from a survey. Transp. Res. Rec. **2542**, 9–16 (2016). https://doi.org/10.3141/2542-02

Docherty, I., Marsden, G., Anable, J.: The governance of smart mobility. Transp. Res. Part A: Policy Pract. **115**, 114–125 (2018). https://doi.org/10.1016/j.tra.2017.09.012

Enoch, M.P.: How a rapid modal convergence into a universal automated taxi service could be the future for local passenger transport. Technol. Anal. Strateg. Manag. **27**, 910–924 (2015). https://doi.org/10.1080/09537325.2015.1024646

Fearnley, N.: Micromobility and urban space. Built Environ. **47**, 437–442 (2021). https://doi.org/10.2148/benv.47.4.437

Fearnley, N., Aarhaug, J.: Subsidising urban and sub-urban transport – distributional impacts. Eur. Transp. Res. Rev. **11**, 49 (2019). https://doi.org/10.1186/s12544-019-0386-0

Fearnley, N., Karlsen, K., Bjørnskau, T.: Elsparkesykler i Norge: Hovedfunn fra spørreundersøkelser høsten 2021, TØI-rapport, 1889/2022, Oslo, Transportøkonomisk institutt (2022a). https://www.toi.no/publikasjoner/elsparkesykler-i-norge-hovedfunn-fra-sporre undersokelser-hosten-2021-article37559-8.html

Fearnley, N., Leiren, M.D., Skollerud, K., Aarhaug, J.: Benefit of measures for universal design in public transport. European Transport Conference/Association for European Transport (AET) (2009)

Fearnley, N., Øksenholt, K.V.: Universell utforming i transportsektoren. NEHF-forlag (2022)

Fearnley, N., Veisten, K., Nielsen, A.F.: Effekter av universell utforming: livskvalitet, etterspørsel og samfunnsøkonomisk nytte. In: Fearnley, N., Øksenholt, K.V. (eds.) Universell Utfroming in transportsektoren. Oslo: NEHF (2022b)

Flügel, S., et al.: Value of travel time and related factors. Technical report, the Norwegian valuation study 2018–2020, TØI-rapport, Oslo, Transportøkonomisk institutt (2020). https://www.toi.no/getfile.php?mmfileid=53108

Fyhri, A., Sundfør, H.B.: Do people who buy e-bikes cycle more? Transp. Res. Part D: Transp. Environ. **86**, 102422 (2020). https://doi.org/10.1016/j.trd.2020.102422

Geels, F., Kemp, R., Dudley, G., Lyons, G.: Automobility in transition?: a socio-technical analysis of sustainable transport. Studies in Sustainability transitions. Routledge, London (2012)

Gregersen, F.A., Flotve, B.L.: Funksjonsnedsettelser - en dybdeanalyse av den nasjonale reisevaneundersøkelsen 2018/19, TØI-rapport, Oslo, Transportøkonomisk institutt (2021)

Haarstad, H., Aarhaug, J., Holm, E.D., Lervåg, L.E., Seehus, R.A., Malmedal, G.: Digitalt skifte for transport - 16 nye teknologier og hvordan de endrer byene, Teknologirådet - Norwegian board of technology (2020). https://teknologiradet.no/wp-content/uploads/sites/105/2020/09/Digitalt-skifte-for-bytransport_endelig.pdf

ITF-OECD. 2018. ITF work on Shared Mobility. https://www.itf-oecd.org/itf-work-shared-mob ility. Accessed 13 Nov 2018

Julsrud, T.E., Farstad, E.: Car sharing and transformations in households travel patterns: insights from emerging proto-practices in Norway. Energy Res. Soc. Sci. **66**, 101497 (2020). https://doi.org/10.1016/j.erss.2020.101497

Kristensen, N.B., et al.: Mobilitet for fremtiden, Transport-, Bygnings- og Boligministeriet (2018)

Lenz, B.: Smart mobility–for all? Gender issues in the context of new mobility concepts. In: Uteng, T.P., Christensen, H.R., Levin, L. (eds.) Gendering Smart Mobilities, 1st edn. Routledge, London (2020)

Lodden, U.B.: Enklere kollektivtilbud. Barrierer mot kollektivbruk og tiltak for et enklere tilbud, TØI-rapport, 540/2001, Oslo, Transportøkonomisk institutt (2001)

Lunke, E.B., Fearnley, N., Aarhaug, J.: Public transport competitiveness vs. the car: impact of relative journey time and service attributes. Res. Transp. Econ. (2021). https://doi.org/10.1016/j.retrec.2021.101098

Lunke, E.B., Fearnley, N., Aarhaug, J.: The geography of public transport competitiveness in thirteen medium sized cities. Environment and Planning B-Urban Analytics and City Science, May 2022 Artn 23998083221100265 (2022). https://doi.org/10.1177/23998083221100265

Mace, R.L.: Universal design in housing. Assist. Technol. **10**, 21–28 (1998)

Negroponte, N., Harrington, R., McKay, S.R., Christian, W.: Being digital. Comput. Phys. **11**, 261–262 (1997)

Nenseth, V., Ciccone, A., Kristensen, N.B.: Societal consequences of automated vehicles - Norwegian scenarios, TØI-report, 1700/2019, Oslo, Institute of Transport Economics (2019)

Nielsen, A.F.: Universal design of public transport systems for people with mental health impairments. In: van der Zwart, J., Bakken, S.M., Hansen, G.K., Støa, E., Wågø, S. (eds.) Proceedings from the 4th Conference on Architecture Research Care & Health. SINTEF akademisk forlag, Trondheim (2021)

Nordbakke, S., Hansson, L.: Mobilitet og velferd blant bevegelseshemmede - bilens rolle, TØI-rapport, 1041/2009, Oslo, Transportøkonomisk institutt (2009)

Nordbakke, S., Schwanen, T.: Transport, unmet activity needs and wellbeing in later life: exploring the links. Transportation **42**(6), 1129–1151 (2015). https://doi.org/10.1007/s11116-014-9558-x

Øksenholt, K.V., Aarhaug, J.: Public transport and people with impairments – exploring non-use of public transport through the case of Oslo, Norway. Disab. Soc. **33**, 1280–1302 (2018). https://doi.org/10.1080/09687599.2018.1481015

Perez, C.: Technological Revolutions and Financial Capital. Edward Elgar Publishing (2003)

Preston, J.: Epilogue: transport policy and social exclusion—some reflections. Transp. Policy **16**, 140–142 (2009)

Preston, J., Rajé, F.: Accessibility, mobility and transport-related social exclusion. J. Transp. Geogr. **15**, 151–160 (2007)

Rogers, E.M.: Diffusion of Innovations. Simon and Schuster (2010/1962)

Schwanen, T., Lucas, K., Akyelken, N., Solsona, D.C., Carrasco, J.-A., Neutens, T.: Rethinking the links between social exclusion and transport disadvantage through the lens of social capital. Transp. Res. Part A: Policy Pract. **74**, 123–135 (2015)

Seehus, R.A., Aarhaug, J., Lervåg, L.E., Haarstad, H., Malmedal, G., Holm, E.D.: Selvkjørende biler - teknologien bak og veien fremover, Oslo, Teknologirådet (2018). https://teknologiradet.no/wp-content/uploads/sites/105/2018/11/Selvkjorende-biler-teknologien-bak-og-veien-fremover.pdf

Sheller, M., Urry, J.: The new mobilities paradigm. Environ. Plan. A **38**, 207–226 (2006)

Skartland, E.-G., Skollerud, K.H.: Universell utforming underveis -en evaluering av universell utforming på bybanen og stamlinjenett for buss i Bergen, TØI-rapport, 1533/2016, Oslo, Transportøkonomisk institutt (2016)

Smith, G., Hensher, D.A.: Towards a framework for mobility-as-a-service policies. Transp. Policy **89**, 54–65 (2020)

Story, M.F., Mueller, J.L., Mace, R.L.: The universal design file: designing for people of all ages and abilities (1998)

United Nations. The convention on the rights of persons with disabilities (2010)

Uteng, T.P., Julsrud, T.E., George, C.: The role of life events and context in type of car share uptake: comparing users of peer-to-peer and cooperative programs in Oslo, Norway. Transp. Res. Part D: Transp. Environ. **71**, 186–206 (2019a). https://doi.org/10.1016/j.trd.2019.01.009

Uteng, T.P., Singh, Y.J., Hagen, O.H.: Social sustainability and transport: making 'smart mobility' socially sustainable. In: Shirazi, M.R., Keivani, R. (eds.) Urban Social Sustainability. Routhledge, New York (2019b)

Veisten, K., et al.: Kollektivtrafikanters verdsetting av universell utforming og komfort, TØI-rapport, Oslo, TØI (2020)

TRIPS: Co-design as a Method for Accessible Design in Transport

Elvia Vasconcelos[1]([⊠]), Kristina Andersen[1], Laura Alčiauskaitė[2], Alexandra König[3], Tally Hatzakis[4], and Carolina Launo[5]

[1] Technical University of Eindhoven, Eindhoven, The Netherlands
{e.m.vasconcelos.de.gouveia,h.k.g.andersen}@tue.nl
[2] ENIL, Brussels, Belgium
laura.alciauskaite@enil.eu, laura.alciauskaite@gmail.com
[3] German Aerospace Center, Institute of Transportation Systems, Berlin, Germany
alexandra.koenig@dlr.de
[4] Trilateral Research, London, UK
tally.hatzakis@trilateralresearch.com
[5] CLIConsulting, Genova, Italy
c.launo@cliconsulting.it

Abstract. Co-design is an established method for ensuring a more democratic approach to design and change propositions. It is however not without friction. In this chapter we describe parts of a process aimed at co-designing inclusive transport systems. In response to the friction of putting such theory into practice, we propose a set of coping mechanisms based on participant feedback. We suggest that such mechanisms have the potential to improve the co-design process beyond this particular case.

1   Introduction

Eighty million European citizens face long-term physical, intellectual or sensory impairment (Eurostat 2019). One in four EU citizens report having a disability, characterized by limitations in performing everyday activities for a period of six months or longer (European 2018). At the same time, the group of disabled persons is very diverse, representing a wide range of access needs, concerns, abilities, and objectives.

Transport is a core aspect of independent living and a key right for everyone. However, there is still a lack of accessible transport vehicles and services (Müller and Meyer 2019), and inaccessible transport and infrastructural barriers consistently prevent persons with disabilities from actively and fully participating in society. As a consequence, people are effectively disabled from accessing job opportunities, education, social and leisure activities.

Even though it is a widely accepted fact that the people concerned should be involved in the research process the number of research projects that provide persons with disabilities the opportunity to participate in research is rather limited. There is a need for persons with disabilities to be "involved as consultants and partners not just as research subjects" (Kitchin 2000). Accordingly, Wilson (2003) recommends: "Disabled people

© The Author(s) 2023
I. Keseru and A. Randhahn (Eds.): *Towards User-Centric Transport in Europe 3*, LNMOB, pp. 173–193, 2023.
https://doi.org/10.1007/978-3-031-26155-8_11

need to be consulted in the design, delivery and implementation of accessible transport systems" (Wilson 2003). By starting a dialogue with vulnerable-to-exclusion citizens and involve hard-to-reach or excluded groups in transport planning the accessibility of transport can be improved.

A way to consider the needs of persons with disabilities in the design of accessible transport is Participatory Design Research. The premise of participatory research is the notion that people are the experts of their own needs and requirements. As a result, participatory research can be seen as a collaborative process that is driven by 'the participation of the persons who will be affected by the output that is being designed.' (Cozza et al. 2020). Thereby, participatory research offers ways for participants to contribute with their expertise and experience, ultimately shaping the research agenda and methods of a project. There are several examples of co-creation projects that involved users in transport planning (e.g., Pappers et al. 2021), however there are few examples of projects that involved persons with disabilities in creating accessible transport (e.g., Gebauer et al. 2010; Luhtala et al. 2020; Tambouris and Tarabanis 2021). As a result, the application of co-design in the context of accessible transport systems remains an open area of research.

In this chapter we describe co-design in the context of accessible transport development, prompted by the friction encountered when the theoretical framework of co-design is put into practise, and we address the following research questions:

- What are the challenges for putting co-design into practice with persons with disabilities to create accessible public transport?
- What coping mechanisms can be put in place to address such frictions, what results do they deliver and how can they be used to improve the experience of doing co-design?

2 Methodological Approach

The work described here takes its origin in TRIPS (www.trips-project.eu), a 3-year European research project aimed at making public transport more accessible for persons with disabilities and elderly travellers. The goal of is to design, describe and demonstrate practical steps to empower people with mobility challenges to play a central role in the design of inclusive digital mobility solutions. In this, we are inspired by Liz Jackson: "You only need empathy in design if you have excluded the people you claim to have empathy for." (Jackson 2019), which in turn can be seen as a reinforcement of the statement "nothing about us without us" (Nothing About Us Without Us). The project-work is done with seven groups of persons with disabilities located in the following European cities: Bologna, Brussels, Cagliari, Sofia, Stockholm, Lisbon, and Zagreb. These city groups are each constituted by a small team of persons with disabilities, tasked with focussing the outcomes of the project to their own ends and needs.

We approach the research questions through the use of observation and interviews with a focus on sketching as a mechanism of conversation and exchange. The outcomes are deeply qualitative in nature and in order to do justice to this, we report in a narrative manner. Our knowledge contribution here is to propose a set of coping mechanisms, co-created in response to challenges encountered, as potential tools to improve co-design in future work.

2.1 A Definition for Working Collaboratively

We understand co-design to be the action of designing together, while actively involving all those implicated in the design process (in this case citizens, government authorities, transport providers and the institutional partners involved in the project) to ensure that outcomes respect all participants' needs and points of view. We found that co-design, co-production, and co-creation were being used interchangeably by the different institutional partners, each coming with different academic backgrounds and specific ways of working. To create a shared understanding of how to work collaboratively, we created the following working definitions: we work under the umbrella of co-production (ethos, attitude, and approach), making use of both co-design (systems, scope, and shared notions) and co-creation (production of explicit design material). We collectively defined them as follows:

We Think of Co-production as the Idea. Co-production generates knowledge in collaborations between people, technology, and society. It is centred on the idea that we can come together in difference and collaboratively create new ideas and concepts. Everyone shares their knowledge, skills, and resources. This also means everyone shares responsibility for making the process successful.

We Think of Co-design as the Action. Co-design describes the action of designing together, while attempting to actively involve all those implicated (citizens, government authorities, transport providers and institutional partners) in the design process to help ensure that outcomes respect all participants' point of view. The aim is to make sure that the process is shared, and the participants feel engaged with the outcomes.

We Think of Co-creation as the Making of Design Material. Co-creation is the act of making together rather than consulting people and then producing designs to the preset requirements. Co-creation involves all actors in the process as active creators of their own futures.

2.2 Participants as Co-researchers

Together we form three groups: participants, researchers, and stakeholders. We acknowledge that these distinctions are constructed, and like any form of classification, they come with a hierarchical logic. However, we put them here in an effort to make the power dynamics at play in the project explicit and make them more addressable in future work.

Participants are the seven local groups of persons with disabilities. Researchers are the people employed to facilitate the project. Stakeholders is every other institutional entity that is engaged in the project such as associations for independent living, as well as local government authorities and transport providers.

Throughout, participants are actively involved in the process as co-researchers and contributors, meaning that their interests are considered drivers throughout the design process. As a first task, each city group defined their goals for the project, and how this vision relates to the overall project. We argue that by making these aims explicit, the participants are constituted as subjects into the project. This is one of our main positions: to involve persons with disabilities in the creation of accessible public transport means constituting them as decision makers in the processes that shape public transport. In this way, we aim to establish a direct parallel between the methods we are using to facilitate the project and the co-design methodology that is one of the main outcomes.

2.3 Participatory Framework

One of the main aims of the project is to guarantee that the people most affected by a change-process are centred in the planning and development of it and are in control of determining what this process is used for, and how it will affect their lives. This is reflected in the framework that demands that participants are actively involved as equal contributors, meaning that their interests are considered valid drivers, and they hold agency and decision-making power throughout the process. Practically, this has been enacted by making explicit the priorities of persons with disabilities, placing them at the centre of our processes - and by shifting our attention towards accounts of what going through this process means from within the participant's social realities.

From this, we take forward that engagement with the structural conditions that exclude people from having access to decision-making processes is fundamental to enable participation. In this we are also inspired by Hamraie's framing of participation, which allows us to use intersectionality as a lens to expand the notion of participation: "It is important to note that participation is not only for persons with disabilities or people with access needs. It precipitates the need for design to understand the experience of the built environment from multiple axes of identity e.g. disability, gender, class, and race (among others), through which more collective, overlapping, and intersectional exclusions can be addressed" (Hamraie 2013). In other words, in TRIPS we are looking to extend the reach of participation towards the more structural framings of decision-making.

Lastly, our methodological stance is firmly grounded in participatory inquiry approaches, where knowledge is generated in a collaborative and iterative manner, and research and action are linked together by critical reflection. This framing is based on established theories and practices from Participatory Design Research (Simonsen and Robertson 2013) (Cozza et al. 2020) (Halskov and Hansen 2015) (Thiollent 2011), Participatory Action Research (Salazar and Huybrechts 2020) (MacDonald 2011) (Hall 1992) (Baum et al. 2006) (Action Research Network of the Americas, n.d.), Research through Design (Andersen and Wakkary 2019) (Frayling 1993) (Giaccardi and Stappers 2017) (DiSalvo et al. 2014), and Design for All (Hamraie 2013) (European Disability Forum, n.d.) (European Commission Employment Social Affairs and Equal Opportunities DG 2010) (United Nations 2006). Grounding our methodological framework in these participatory traditions, allows us to create common ground and understanding between persons with disabilities and institutional actors in the TRIPS project, nurturing collaborative processes that make mobility concerns and concepts visible, while integrating cultural, interpersonal, structural, and policy-related viewpoints.

In the following, we describe the methods (Fig. 1) currently being co-designed with the seven cities. We report on the lived experience of making use of these methods, specifically on the challenges encountered with putting these ideas of into practise and the coping mechanisms we have produced to make things work.

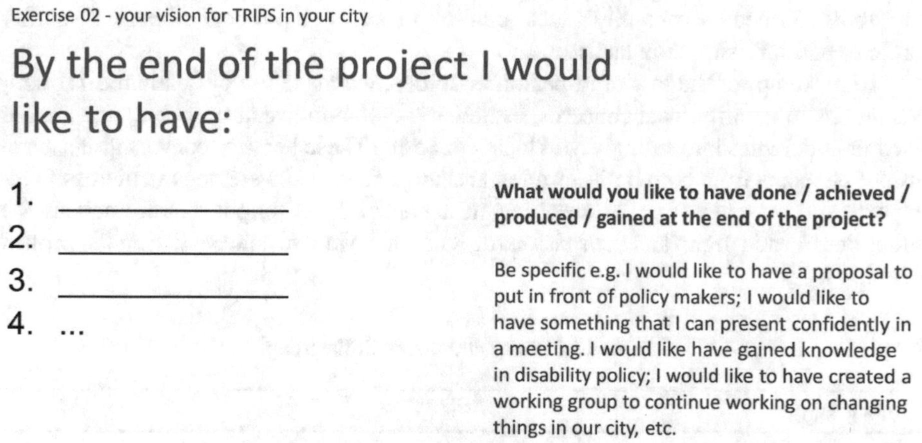

Exercise 02 - your vision for TRIPS in your city

By the end of the project I would like to have:

1. _____

2. _____

3. _____

4. ...

What would you like to have done / achieved / produced / gained at the end of the project?

Be specific e.g. I would like to have a proposal to put in front of policy makers; I would like to have something that I can present confidently in a meeting. I would like have gained knowledge in disability policy; I would like to have created a working group to continue working on changing things in our city, etc.

Fig. 1. Slide from workshop method

2.4 Co-design Fieldwork

The TRIPS methodology development work was initially intended to be executed through a string of in-person activities allowing the approach to be designed in bursts of iterative sessions in each city. However, since the entirety of the project has been conducted online due to Covid-19 restrictions, our work has taken on a much more elaborate and localised form.

Practically, we planned to make use of four main techniques:

Workshopping: Through workshopping we aim to create an experience where individuals' narratives coexist with complex understandings of collective knowledge, leading to a great diversity in outcomes.

Brainstorming: Brainstorming allows for a broad range of knowledge to manifest, be shared and co-created. This has a dual effect in user involvement: it generates possibilities and equally improves the social dynamics of exchange as a basis for shared meaning.

Sketching: Through sketching we aim to explore notions of collaborative visual thinking in which nonverbal techniques like drawing are used to represent unified action.

Interviews: Interviews elicit individual knowledge and narratives. We propose to use them as open engagements where personal stories guide participants and interviewers in the telling of lived experience.

For the practical purpose of working within Covid-19 restrictions, these methods were re-purposed to be executed online and in smaller groups. Each local city group had to develop a way to work together locally and with the overall project, considering the limitations of online platforms. Our role as researchers became to facilitate the work in a way that built on the specificities of each local team in terms of knowledge, skillset, and tools, framed by each city's local context in terms of political climate, disability policies and infrastructural landscape.

To make up for the loss of in-person activities, we made use of continued 1:1 conversations to identify local concerns, establish a collaborative atmosphere, and anchor the methodologies into strongly held local concerns. These iterative conversations made use of elements from both brainstorming and interviews and were documented as field-reports and sketches (Fig. 2), resulting in a consolidated output that formalises the identities of each group and their vision for what they want to achieve during the project (Table 1).

Table 1. Identity and vision Sofia group

Sofia group	
Who are we	The Sofia group is constituted by persons with disabilities who are working together to make public transport in Sofia accessible, safe and comfortable Our efforts are grounded in the present and we are focused on the problems we are faced with right now when using public transport. Our starting point is a common problem: bus stops in Sofia are not accessible
Our vision	Make public transport in Sofia accessible, safe and comfortable Solution: start with bus/trams stops and metro stations We want to see change happen in Sofia in the duration of TRIPS. Our efforts are guided by knowing that: 1. The change we want to see in Sofia has to do with practical solutions we would like to implement in our city to address the problems we have today. This is a common problem that we think can be addressable within the duration of TRIPS: bus stops are not accessible 2. Creating common ground between persons with disabilities, local authorities and other institutional partners is crucial to turning our vision into something that can be practically achieved

In this process, we focused on the localisation of the work based on the notion that in order to create meaningful change, processes must be anchored in deeply felt concerns. Our expanded notion of participation is grounded in people's ability to determine and shape the environment of their everyday lives. We extend this principle to our own processes: participants determine and shape the conditions of their own participation in the project and the extent to which this affects their lives. This can be seen in the decision by the Sofia group to stay on the practical goal of the bus stop (Table 1).

This work was conducted through a combination of qualitative research methods: semi-structured interviews, open-ended activities, writing exercises, surveys, offline activities, etc. Our focus here has been to set a dynamic working rhythm and generate mechanisms that allow heterogeneous interests and in-depth understandings to come forward. As such, we prioritise research that produces knowledge grounded in the everyday and that stems from the realities of the local communities we are working with. This means that our data collection has mostly gathered descriptive insights, that explain what is happening in detail and how something is experienced from the perspectives of the local city groups.

As a result, our research approach prioritises first-person, subjective perspectives and generates descriptive and rich insight. This type of qualitative outcome supports an in-depth understanding of the situation being studied and generates knowledge that is locally situated in each city and is specific to each group. Our developing thesis is that such local concerns can be explored in depth by each city group and then shared to other locales, together forming a series of complimentary exemplars of the methods-in-action. This work then forms the experimental backbone of the methodology work in the project, allowing the developing ways of working to be explored in specific situations, anchored by the concerns and commitments of each group, and through that validate methods and explore the extent that techniques can be transferred between different cities, situations, and concerns.

In the following, we will re-narrate a set of challenges we encountered in this work as a way to highlight relevant insights and open them for the discussion. In doing so, we follow Maria Puig de La Bellacasa's notion of "thinking with care" (Bellacasa 2017), together with Nigel Rapport's description of interviews as a "form of partnership" and "an extraordinary encounter" (Raport 2012) and Arthur Frank in his proposal to make use of dialogical narrative as a way of letting "stories breathe" (Frank 2010). We take these three theoretical positions to inspire how we process the outcomes of the ongoing conversations with the cities.

3 The Challenges of Putting Co-design into Practise

Having secured funding with a proposal built on extensive levels of co-design, we immediately met a series of challenges as we started putting these ideals into practise. These challenges were split into three groups: expectations unmet; diverging methods and needs of the participating groups; and finally, the practical realities of collaborating online in the context of COVID-19. In the following, we identify these challenges and describe the coping mechanisms we used to continue working through, with and alongside these problems that are both difficult and intrinsically unsolvable. Throughout the text, we quote from interviews and feedback sessions with participants and partners.

3.1 Expectations Unmet: Arriving. Onboarding. What Do You Bring with You into This Space?

Following the initial onboarding period, all city groups started engaging with the different tasks determined in the project's roadmap. "People got engaged in a very sincere way,

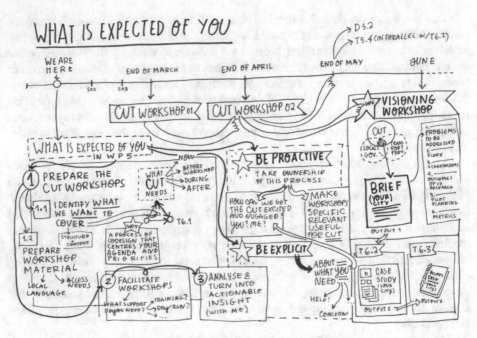

Fig. 2. Sketch used in 1-2-1 session with group

because they want to make a change." After a couple of months, some participants started voicing dissatisfaction with the ways they were being engaged with. "We are just told, next week you need to do this."

The dissatisfaction came both from small and big things: not being consulted with enough notice, not having full visibility of tasks, not being part of the planning. "You can't tell me to attend a workshop on a Saturday the Wednesday before - I have a busy life and we need to be consulted about these decisions." These frustrations started to surface the underlying assumptions that each group had about what working in a co-design manner would be like, and how they needed their varying expectations to be met.

3.2 Diverging Methods and Needs: Settling In. Variation. How Can You Coordinate Parts of an Ever-Changing Whole?

"This is ok, but it's not co-production…Just call it a workshop." As the project unfolded, different methodological approaches started to surface. Not only did we see different methodologies at play, but we observed that different groups meant different things, when they used words such as participation and co-design. At the same time, it became clear that each group had unique challenges and opportunities in the project. The groups were not only different in terms of age profile, gender distribution and types of disabilities, but also in experience, interests, motivation and focus. We also had to consider each group's unique contextual frame of local disability policy, transport ecosystems, accessibility culture, and the participation of persons with disabilities in public decision making.

This meant that the project was understood and experienced very differently by the different city groups, and this created tensions between the project's timeline, and what was considered relevant and urgent in each location. These tensions manifested the need for a more flexible approach to the project's predefined structure, to allow the cities to work in the most meaningful way given their local situation. "Don't speak about flying taxis without speaking about accessible bus stops first." While the gap between what was expected and what could be delivered, grew in some cities, other cities were able to take ownership of tasks to make them meaningful to their needs: "We separated the workshops and created a separate one with just the experts. It was relaxed and we had time to go through things thoroughly. People were very interested, there was an active discussion."

3.3 Collaborating Online: Teams. Zoom. Skype. Google Drive. Whatsapp. Slack. Can You Share that Document?

"Here's the group case study, I have no idea how to add it to the drive or where that is." Finally, we were challenged by the fact that we all went into lock-down as soon as the project started, and what was supposed to be an in-person methodology had to be pulled online. This meant that we suddenly needed a high degree of technical literacy, and our work became dependent on existing online platforms and written language (in English). We rapidly re-purposed our methods to be executed online, experimented with platforms, and started working in much smaller online groups.

These online tools brought some advantages: it was easier to stay in touch without travelling and freed from the requirements to travel, some participants were able to do more. We see the irony in this. "The airline cancelled my flight for no reason." These online interfaces also came with a whole different set of disadvantages and accessibility issues: it became clear that creating engaging activities online was much harder and the potential for misalignment was greater.

"You are muted." We made use of online meetings and shared documents to work with each group. However, setting up collaborative ways of working online was a surprisingly hard task. We found it hard to share information, to keep documents up to date, to navigate folder structures without getting lost and we also observed that it can be difficult to ask for help.

4 Coping Mechanisms: Survival

To address the challenges encountered, we had to produce other ways for making things work with the cities. These simple coping mechanisms were emergent as fixes and compensations, and in themselves they relied on goodwill and a re-address of the aims of the project. In the following, we describe the ways we coped with the challenges by engaging in a set of smaller, discrete action points: listening; nurturing local variation; integrating multiple methods; setting the agenda; and digital skills.

4.1 Listening

The first coping mechanism came from a moment of reflection, where we took a step back from the plan, to pause, listen and pay attention to what was at play in each of the groups outside the immediate task at hand. As researchers, we felt conflicted about pausing. On the one hand, we wanted to address the issues being brought forward by the groups, on the other hand we feared that giving attention to issues that went beyond the scope of the task could lead the groups too far away from the work. In short, we had a map with a route, and we felt that as researchers our role was to guarantee we did not get lost, and nobody was left behind.

Through listening we found that each city arrived at tasks on very different footings and subsequently, that they continued to travel their own unique paths. In contrast, researchers reported they often found themselves working very hard to 'herd' the groups into a shared and simultaneous path, to navigate the task and stick to the plan. By taking a moment to listen, we were able to challenge the need for one/single/right/sequential order of work. We started to make space for variation in our methods. This required us to stop forcing a coming together, towards thinking about collaboration as a kind of social choreography, as the coordinating of similar and simultaneous gestures (Hlavajova 2020) across distances (literal or figurative) but not necessarily at the same time. And so, we folded the map, and started taking the lead from each of the city groups, following them in their own unique paths.

4.2 Nurturing Local Variation

Through this experience we came to acknowledge that co-design can mean different things in different contexts, and therefore we also came to recognise that there are many different ways of doing co-design. As a result, the TRIPS co-design methods are developed as a way of working that emerges from within each of the cities involved, and this requires us to make space for plurality to unfold. This is particularly evident in the ways each of the seven cities have developed their own paths in the project. As a result, our methods have been continuously adjusted to nurture local variation between cities and build on the strengths of those involved (Fig. 3).

To nurture local variation in our processes, we proposed activities that were doable in ways that were grounded in the experiences of each group. To support this notion, our methods were crafted around four high level categories of variation that we saw manifesting in the cities: people, setup, place, and time. These were broken down into specific variants such as whether there was an existing working relation between the people involved. We found that not all variants had the same weight in terms of how they impacted the ways each city engaged in the project. Through this we were able to identify the variants that embodied a significant impact: (1) whether the group was led by a disabled person; (2) the contractual setup; and (3) motivation and expectations.

Through local variation, the project acquired multiple trajectories that were continuously re-shaped as each city navigated the project. To support these multiple trajectories, we found it necessary to develop mechanisms to make our processes flexible, but also tight in order to create a more seamless experience to join up the discrete components of the project. These methods acknowledged that each city started and continued to develop differently as the project evolved. This required different responses from the project that needed to be quick and persistent in adjusting the work to follow and support what was meaningful and possible on a city-by-city basis.

4.3 Integration Multiple Methods

To cope with the different trajectories emerging in each city, we became particularly focused on generating the mechanisms to allow for multiple insights to come forward, develop and settle. We worked with the understanding that co-design does not mean that everyone is the same, rather it means that everybody is meeting in their differences. We found it necessary to also extend this notion of plurality to the internal ways of working in the project: not all parts of the project were set up to be co-designed, nor were they expected to. However, we found it necessary to reconcile this with the understanding that the different methodological approaches at play in the project were impacting the experience of the city groups. Out of this understanding, came a need for ways to integrate multiple methods to work both in parallel and as a whole. Our efforts were guided by knowing "that these approaches need not be mutually exclusive (…) it is possible to include models that are radically different and to allow multiple models to coexist—separately or layered or even integrated with each other" (Olson 2007).

4.4 Setting the Agenda

Another coping mechanism that emerged out of the unique trajectories of each city, proposed to make use of making as a way to broaden participation. To set up participation towards variation and difference, rather than sameness, we used the making of things to support individual structures of knowledge to emerge.

As the project progressed, we observed that the seven groups of persons with disabilities were the only stakeholder group in the project without a formalised identity. All other stakeholders were institutions that came into the project with their identities and agendas fully formed. The lack of a defined identity started to manifest in the interactions between the city groups and the actors in their local transport ecosystems. "We keep on explaining to people why we started this project: we want to see co-production in making transport disability friendly - how to make this more concrete and practically what does this mean?".

The city groups were struggling to communicate clearly who they were, what they were trying to do and why people in their cities should be interested. This became an obstacle to making connections with government authorities and transport providers, a fundamental component of the project. To cope with these issues each city was engaged in the making of a number of artefacts to define and communicate their specific motivations and goals. An important artefact coming from this work was the identity and vision document that outlined in a simple and clear way, who each group was, and what they set out to achieve in the duration of the project.

Through the making of these artefacts the cities achieved two things: first they positioned each group's agenda as a driving priority in the project, effectively placing the problems these groups experienced with public transport at the centre of our efforts; and second, these artefacts supported the groups in establishing the much-needed connections with the municipalities and transport providers in the cities. Further, through the making of things the groups were able to actively influence the direction of the project. This positioning was sustained in our methods by paying attention to the specific areas of interests articulated in each city. This required us to redirect our efforts to support the cities in their work towards tangible results whilst being realistic about the scope for change that could be achieved in the duration of the project (Fig. 3).

4.5 Digital Skills

To cope with the demands of online work we were forced to place a significant amount of effort to make our collaborative processes work online.

In this process we found that a digital setup did not reflect the ways most of the city groups involved in the project were used to working and this posed varying levels of challenges to turn online spaces into true shared environments. We also faced difficulties with setting up collaborative processes with the multitude of organisations involved in the project, considering local platform preferences, as well as security concerns and storage. In these efforts, we identified varying levels of digital skills that we addressed by working on a 1:1 basis to share knowledge and train people in using these platforms. We used a combination of platforms and several shared documents to cater for the specific needs of each city. An example of this is the access needs protocol that explains the practical setup each city group needs to participate in an online session, something that has been used internally in the project as well externally, to engage with the local stakeholders in each city. As more documents piled up, we doubled our efforts to keep information easy to access and actionable.

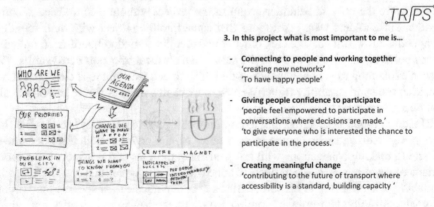

Fig. 3. Slide with quotes from the groups about the project

5 Protocols and Templates as Methods

The artefacts and documents created in the project form the main site for the co-design of our methods. The goal for the co-design methodology-for-all is to produce a set of working templates and protocols that aid participants in the creation of an identity, vision, goals, and approach. As such we take inspiration from design research methods that generate use cases and areas of concern as well as business and entrepreneurship canvasses that identify, visualise, and assess an idea or concept.

"Protocol. A set of principles to work by some guidelines for the guests, so you don't have to reinvent the wheel time and time again. A structured daily rhythm of rituals, helping to move forwards." (Gysel 2018).

Through the process of working together, a set of documents, templates and slides has evolved into working protocols for generating scope and aims for inclusive transport projects. These documents have been used as living documents, continuously evolved to cope with the emerging needs of each group. We found that professional looking documents were needed to create a sense of confidence and validation of the work we were doing (Fig. 4), and at the same time, the use of more quick and dirty materials such as draft sketches (Fig. 3) were a productive way to break down theory and business jargon into something that is understandable, relatable, and motivating.

These materials are practical in nature and aim to support the groups in articulating their work (Table 2). As an example, the Access Needs Protocol was created to communicate the practical setup that each group needed to participate in an online session: "Two members of the working group are blind. One of our members requires an interpreter to participate." We were inspired by Sandra Lange's 'Access Rider Exercise' (Lange) in prompting each group to create and inhabit a shared space (online and physical) whilst actively shaping the conditions of the interaction in that space. Like all other protocols, the Access Needs Protocol has a function beyond the project, in this specific case the intention is to promote the equal participation of the groups in public spaces such as meetings with the city council. Through these co-developed documents, we have also extended the cities' collaborative and situated practises to the multiple partners involved.

These protocols and methods have been continuously developed to reflect the ongoing and emerging thinking of each group as they articulate their work in the project. They are both the conditions and part of the outputs of putting co-design into practise.

5.1 Reflecting on Our Working Definition and Aims

We shared our initial definitions for co-production, co-design, and co-creation earlier, in order to explain how these concepts work together and how knowledge can be generated. They also set the principles and ground rules for the working practises in the project. As our collaboration unfolded, these definitions were re-visited to reflect on our learning and the experiences of each city group in the project (Fig. 5). We now understand that for TRIPS, these terms hold multiple definitions that provide living accounts of the localised processes created by each city. Below we share our current understanding of these terms.

- **Co-production in TRIPS** is about whose voices are heard, emphasising that the voices of persons with disabilities need to be heard louder. Co-production is doing

Table 2. Overview of workshop design method

Workshop method	
Step 01	Define the purpose of the workshop Who is this workshop for and what for? What do you wish to have as an output of this workshop?
Step 02	Create an agenda Start a high-level agenda by splitting your time into blocks. Add as you go along
Step 03	Produce workshop materials Linking to templates (unpacked below)
Step 04	Define the practical setup This entails putting to practise the access needs protocol established with your group, defining roles and responsibilities e.g. timekeeping, technical troubleshooting, together with deciding what information needs to be sent in advance e.g. email reminder -24 h in advance, and setting up the documents to be used during the workshop e.g. collaborative note taking
Step 05	Turning insight into actionable outputs Straight after the workshop do a quick 15 m debrief with all facilitators. Start with a 5 m silent brain dump: what are my 3x main takeaways? What do I think we need to do next? Any urgent critical things? Discuss and cluster After a day or two go through the notes and summarise the main insight coming from the session. Discuss these with the other facilitators

Pilot City	Brussels						
The reference local coordinator (LUL) from the disabled Community	Fabrizio Albani						
The reference local coordinator (LC) from the local stakeholders Community							
Inclusive Digital Mobility Solution to be developed	**Name/Title** Accessible Journey Planner Based on user profile and type of deficience						
Description	Accessibility Journey Planner Application/website for disabled persons • Reinforcement of Brussels Online Accessibility mapping information in the STIB's *Trip Planner* • Study of current STIB's *Trip Planner* • Informs about the accessibility of the journey based on user profile/type of deficience (visual, hearing, physical, mental, intellectual) • Reinforcement of the Accessibility of the STIB's *Trip Planner* website/application itself Option in the app: Contact Assistance or Alternative transports in case of problem during the journey (Stib Contact Center and Assistance, STIB Taxibuses contact, other adapted transports contacts (Taxis Verts, …) Option in the app: possibility to report (by sending a picture) an accessibility problem or dysfunction						
Transport Service Involved	Railway	Bus	Metro	Taxi	Taxibus	Car	Bike
	x	x	x	x	x		

Fig. 4. Pilot case study template

things together on an equal and meaningful basis and is reliant on having a common vision that builds on the strengths of everyone involved in the process.

In practise co-production means working processes that emerge from within, that are firmly grounded in the realities and the specificities of the local ecosystem of each city. This means each group explores their own path and makes the process their own. Special attention needs to be paid to motivation as fundamental to the success of any working process.

- **Co-design in TRIPS** looks like specific and tangible actions with results that reflect the conversations they come from. Participating groups want to know they are being listened to, and that their work is contributing to change. It is an end-to-end process of exchange and an ongoing practise that requires a multitude of working processes. Co-design looks like having ownership of the process and being fully engaged in shaping it.

In practise co-design means producing locally informed responses to localized challenges (Todd 2015), together. This entails embracing variation through acknowledging the contextual specificities unique to each group whilst nurturing what they share: a drive to improve public transport for persons with disabilities in their cities and an appetite to learn and connect to the other cities in the project. Codesign means listening from the very beginning all the way to the very end.

- **Co-creation in TRIPS** is centred on taking coping mechanisms and turning them into protocols, that go beyond the immediate situation. This means establishing the common ground where people can connect and learn from how each group is working through the process and barriers so that ultimately, they can affect change in their cities.

In practise co-creation means creating meaningful change in whichever way is meaningful to people. This entails enabling individual goals, supporting the work of local communities, and nurturing productive synergies between cities. The outcome is bringing all the moving parts together to create concrete change.

6   Findings and Discussion

In effect, we are co-designing the way we collaborate, as we work. This leads to an emerging set of insights, methods, and strategies for co-designing ownership of our processes. This discussion takes a second look at the coping mechanisms generated with the cities. The aim is to go beyond the survival functions in these mechanisms, towards something that may be used productively and strategically to mobilise this kind of project going forward.

The challenges we encountered manifested a gap between the idealised plan and the realities of doing co-design in practise. Exploring this gap between theory and practise led us to a better understanding of how these divisions are specifically enacted in our methods. Taken further, we believe this led us to understanding the kinds of relations these divisions support, such as the reproduction of asymmetrical power relations between persons with disabilities and stakeholders. We want to signal this, while at the

2 . For me, doing codesign for real looks like

- **Specific and tangible actions**
  'with results that are specific and tangible and reflect
  the conversation that they came from.'

- **An end-to-end guided process of exchange**
  'having everyone from the very beginning and
  shaping together the future/activity/action.'

- **Ownership of the process**
  'Through several methodologies develop the project
  from scratch with the ones who will use it.'

Fig. 5. Slide with definitions of codesign coming from the groups

same time acknowledge that a thorough unpacking of this is beyond the scope of this chapter. What we propose instead is to focus on a smaller subset, that starts with how the coping mechanisms bridge the gap between theory and practise, and the movements this generates.

Beyond acting as an immediate response to challenges, the coping mechanisms both delineate and attempt to bridge the gaps between the theory and practise in a way that generates movement in both directions. This has allowed us to revisit the idealised project plan in light of the necessities and possibilities emerging in each city - a movement (back) from practise to theory. But they also allowed us to shift the efforts in the project towards the ways each group found most meaningful to work in practise - a movement away from theory towards practise. We believe that these movements create productive ways to go forward that prioritise the cities' collaborative and situated practises inside the project.

We saw this particularly at play in the multiple methods and variation setup, which is being extended beyond a survival technique, to a feature in the entire project. We started to make space for variation in our methods as a way to cope with the tensions created by the structure of the project, which sometimes lacked the flexibility to accommodate for each city's local situation. As a coping mechanism, nurturing variation in our methods effectively redirected our efforts to the specific and practical realities of the cities. This expanded the borders of the project to allow for different methodological approaches to coexist, nurture situated practises in the cities and allow for multiple insights to come forward. Extending this further has led us to challenge the entire sequential order predefined in the project. Challenging the need for a rigid sequential order in turn opened up for a shift in the type of knowledge that was prioritised in our processes. As a result, we became better equipped to generate the type of knowledge that comes from "a critical reading of experiences" (Salazar and Huybrechts 2020), which implies a shift in the order of how research is conducted so that practise precedes theory and not the other way around.

This letting-go of a sequential order of project tasks became the door that opened the way for us to consider alternative project structures, such as modular tracks and multiple trajectories. This is an example of a coping mechanism turned strategy that we take with us into future work. We believe that it can be extended to future projects and contribute to wider discussions on how to productively navigate the tensions that emerge in the process of weaving practise and research together in co-production (Chambers et al. 2022).

We are also committing to the movements generated through the 'setting the agenda' mechanism. This is an important mechanism to extend because it opens the way for more strategic moves to continue to support the cities in achieving concrete change. This is particularly necessary as we deal with multiple and varied understandings of what constitutes change. We are seeing both very broad understandings of change, "I would like to have a more accessible city," but also more specific accounts: "It's important to make institutions and municipalities understand that accessibility is a very important thing; that people with disabilities are citizens who have rights, they only want to have a normal life and have the possibility to have a full life, with autonomy and so on." We believe that this move from abstract to specific, is key to refining our methodologies in the future.

On our road towards achieving concrete and tangible change, we look at the role material artefacts play in enabling participation (Noronha et al. 2020). This is supported by thinking widely about the socio-material arrangements that constitute material participation (Marres 2012), and by looking at the practical ways of generating knowledge through the making of things (Giaccardi and Stappers 2017). The emphasis here is placed on what value is generated for the cities and how this can be brought forward through participants' own accounts of how this manifest in their material realities. In these efforts, we look at the work of Grada Kilomba (Kilomba 2020) to find ways to prioritise the groups' subjective accounts of what going through this co-design process means and the value it generates in their cities. In this we are also inspired by Mariolga Reyes Cruz who in 'What If I just cite Graciela?' (Reyes Cruz 2008) explores ways of treating participants beyond data towards constituting them as the theoretical grounding upon which research can be done. We combine this with the work of Ann Light about the situated and interpretative nature of account-making, to explore new models of authorship "to legitimate new practices of feeling, telling and accounting for." (Light 2018).

Ultimately, we are concerned with giving each city a sense of accomplishment and leaving them with ways to continue working towards long term impact.

6.1 Long Term Impact

One way to create long time impact will be to make explicit the sustainability and longitudinal impact of these types of projects. We propose that the methods we are developing might be used in broader contexts of policy development. At the same time, we have become increasingly aware of a structural lack in the context of the European research community. Projects addressing accessibility come and go, but the generated results are often lost or only partly reported, and the disability community is experiencing that they have to start from scratch, again and again. This compounds to the existing accessibility skills gap within the industry, and ultimately to the barriers to structural change.

A way to address this in the future would be through the establishment of a European Accessible Design centre as a platform for addressing the structural lack about the generated results of projects addressing accessibility. Our starting point is an engagement with the structural conditions that exclude people from having access to decision-making processes as a fundamental aspect of what constitutes participation (Hamraie 2013).

"We have realized that advocating for "more persons with disabilities in design" without advocating structural changes to what design is, how it operates, and what problems it seeks to solve is just advocating for a select few people to gain more power within an unjust system, while allowing the marginalization of others by that system to become more entrenched." (Jackson and Haagaard 2022).

As researchers we also reflect on how to formalise the relations created between civil society and institutional actors in our methods and more broadly as a political matter: how persons with disabilities are positioned in each of these cities to begin with, the relations they had, the ones they created, and the ones they failed to build and why.

We believe that a European centre could integrate and progress aspects of accessibility design (process, tools, skills, and overall state of the art), as a hyperlocal, network structure with common organisational and support processes, but also local branches networked with local industry, disability NGOs, communities, and individuals. Allowing the work to be local in its scope, based on deeply felt local concerns with an intimate understanding of contextual, legal, and political barriers to change. We believe that this might work towards the necessity of a social justice orientation in this kind of research work.

7 Conclusion

In this chapter, we have reported on the process of transitioning a project from idealised application to actualised practise. We have described the emerging difficulties of our initial co-design processes and a set of coping mechanisms developed to mitigate the challenges encountered. We consider these mechanisms to be useful beyond their solutionist and survival functions and we propose that they can be used productively and strategically as methods to improve future co-design projects.

Our immediate future contributions will come in the form of a co-design methodology that can be adapted to engage persons with disabilities in decision-making processes for public transport. In this, we seek to establish parallels between co-creating a design process and the decision-making processes relating to public transport involving civil society, transport providers and local authorities. In other words, in constituting each group as equal contributors in the project we're suggesting ways of constituting persons with disabilities as subjects in the context of public transport in their cities. These processes themselves will be co-created, facilitated, and co-owned by the participating cities.

Our contribution here is to report from the co-design process in practice, to describe a set of coping mechanisms formulated as co-design strategies and begin the future work of moving them towards generative tools. In doing this, we have arrived at the understanding that survival techniques can also be used productively and strategically to address more structural issues, both in the present and as a step towards generating long term impact. We believe that this understanding will guide our work in this and future projects.

Acknowledgement. The TRIPS project is made possible by funding from the European Union's Horizon 2020 Research and Innovation Programme Under Grant Agreement no. 875588.

References

Action Research Network of the Americas. Accessed August 2019. https://arnawebsite.org/act ion-research/

Andersen, K., Wakkary, R.: The magic machine workshops: making personal design knowledge. In: CHI Conference on Human Factors in Computing Systems - Proceedings. New York: Association for Computing Machinery, pp. 1–13 (2019). https://doi.org/10.1145/3290605.330 0342

Baum, F., MacDougall, C., Smith, D.: Participatory action research. J. Epidemiol. Commun. Health **60**, 854–857 (2006). https://doi.org/10.1136/jech.2004.028662

de La Bellacasa, M.P.: Speculative Ethics in More than Human Worlds. University of Minnesota Press, Minneapolis (2017)

Chambers, J.M., et al.: Co-productive agility and four collaborative pathways to sustainability transformations. Global Environ. Change **72**, 102422 (2022). https://doi.org/10.1016/j.gloenv cha.2021.102422

Cozza, M., Cusinato, A., Philippopoulos-Mihalopoulos, A.: Atmosphere in participatory design. Sci. Cult. **29**, 269–292 (2020). https://doi.org/10.1080/09505431.2019.1681952

DiSalvo, C., Lukens, J., Lodato, T., Jenkins, T., Kim, T.: Making public things: how HCI design can express matters of concern. In: SIGCHI Conference on Human Factors in Computing Systems. Association for Computing Machinery, New York, pp. 2397–2496 (2014). https:// doi.org/10.1145/2556288.2557359

European Commission. 1 in 4 people in the EU have a long-term disability (2018). https://ec.eur opa.eu/eurostat/web/products-eurostat-news/-/EDN-20181203-1. Accessed 03 Dec 2021

European Commission Employment Social Affairs and Equal Opportunities DG. M/473 Standard-isation mandate to CEN, CENELEC and ETSI to include "Design for All" in relevant standardi-sation initiatives (2010). https://ec.europa.eu/growth/tools-databases/mandates/index.cfm?fus eaction=search.detail&id=461. Accessed Sept 2020

European Disability Forum. n.d. Design for all. Accessed 11 Sept 2020. http://www.edf-feph.org/ design-all

Eurostat. Disability statistics (2019). https://ec.europa.eu/eurostat/statistics-explained/index.php? title=Disability_statistics

Frank, A.W.: Letting Stories Breathe: A Socio-Narratology. The University of Chicago Press, Chicago (2010)

Frayling, C.: Research in Art and Design. Royal College of Art Research Papers (Royal College of Art) 1 (1993)

Gebauer, H., Johnson, M., Enquist, B.: Value co-creation as a determinant of success in public transport services: a study of the swiss federal railway operator (SBB). Manag. Serv. Q. Int. J. (2010)

Giaccardi, E., Stappers, P.J.: Research through design. In: The Encyclopedia of Human-Computer Interaction, pp. 1–81. Interaction Design Foundation (2017)

Gysel, J.: Makers of their own time. GIRLS LIKE, US no. 11. Jessica Gysel (2018)

Hall, B.: From margins to center? The development and purpose of participatory research. Am. Sociol. **23**, 15–28 (1992). https://doi.org/10.1007/BF02691928

Halskov, K., Hansen, N.B.: The diversity of participatory design research practice at PDC 2002–2012. Int. J. Hum.-Comput. Stud. **74**, 81–92 (2015). https://doi.org/10.1016/j.ijhcs.2014.09.003

Hamraie, A.: Designing collective access: a feminist disability theory of universal design. Disabil. Stud. Q. (The Ohio State University Libraries) **33**(4) (2013). https://doi.org/10.18061/dsq.v33i4.3871

Hlavajova, M.: Online Course: How to Assemble Now. BAK Public Studies (2020)

Jackson, L.: Empathy reifies disability stigma (2019). https://interaction19.ixda.org/program/key note--liz-jackson/. Accessed Sept 2020

Jackson, L., Haagaard, A.: The against list. Medium 28 February 2022. https://medium.com/the-disabled-list/the-against-list-64676cb2acc4. Accessed 11 March 2022

Kilomba, G.: Plantation Memories: Episodes of Everyday Racism, 6. Unrast-Verlag, Münster (2020)

Kitchin, R.: The researched opinions on research: disabled people and disability research. Disabil. Soc. **15**(1), 25–47 (2000). https://doi.org/10.1080/09687590025757

Lange, S.: Access Rider Exercise. Accessed 26 Sept 2020. https://docs.google.com/document/d/19iBfrb1GJlEGSPPInt3OyU5rk8dCvcCMeVyi50SHUY0/edit

Light, A.: Writing PD: accounting for socially-engaged research. In: PDC 2018: Proceedings of the 15th Participatory Design Conference: Short Papers, Situated Actions, Workshops and Tutorial. Association for Computing Machinery, New York, pp. 1–5 (2018). https://doi.org/10.1145/3210604.3210615

Luhtala, M., Federley, M., Kuusisto, O., Pihlajamaa, O., Kostiainen, J.: Co-creating mobility services with disabled and elderly people. In: 8th Transport Research Arena, TRA 2020-Conference cancelled, p. 270. Liikenne-ja viestintävirasto Traficom, April 2020

Müller, B., Meyer, G. (eds.): Towards user-centric transport in Europe. LNM, Springer, Cham (2019). https://doi.org/10.1007/978-3-319-99756-8

MacDonald, C.: Understanding participatory action research: a qualitative research methodology option. Can. J. Action Res. **13**(2), 34–50 (2011). https://doi.org/10.33524/cjar.v13i2.37

Marres, N.: Material Participation: Technology, the Environment and Everyday Publics. Palgrave Macmillan, London (2012)

Noronha, R., Aboud, C., Portela, R.: Design by means of anthropology towards participation practices: designers and craftswomen making Things in Maranhão (BR). In: 16th Participatory Design Conference 2020 - Participation(s) Otherwise. Association for Computing Machinery (2020). https://doi.org/10.1145/3385010.3385015

Nothing about us without us. Wikipedia. Accessed Sept 2020. https://en.wikipedia.org/wiki/Not hing_About_Us_Without_Us

Olson, H.A.: How we construct subjects: a feminist analysis. Library Trends, special Issue: Gender Issues in Information Needs and Services (Johns Hopkins University Press), vol. 56, no. 2, pp. 509–541 (2007). https://doi.org/10.1353/lib.2008.0007

Raport, N.: The interview as a form of talking-partnership: dialectical, focussed, ambiguous, special. In: The Interview: An Ethnographic Approach, by Jonathan Skinner. Routledge (2012)

Pappers, J., Keserü, I., Macharis, C.: Participatory evaluation in transport planning: the application of multi-actor multi-criteria analysis in co-creation to solve mobility problems in Brussels. In: Transport in Human Scale Cities. Edward Elgar Publishing (2021)

Reyes Cruz, M.: What if I just cite Graciela? Working toward decolonizing knowledge through a critical ethnography. Qual. Inq. (Sage J.) **14**(4), 651–658 (2008). https://doi.org/10.1177/1077800408314346

Salazar, P.C, Huybrechts, L.: PD otherwise will be pluriversal (or it won't be). In: Proceedings of the 16th Participatory Design Conference 2020 - Participation(s) Otherwise, vol. 1, pp. 107–115. ACM, Manizales Colombia (2020). https://doi.org/10.1145/3385010.3385027

Simonsen, J., Robertson, T.: Routledge International Handbook of Participatory Design. Routledge, Milton Park (2013)

Tambouris, E., Tarabanis, K.A.: Integrated public service co-creation: objectives, methods and pilots of inGov project. In: EGOV-CeDEM-ePart-*, pp. 83–90 (2021)

Thiollent, M.: Action research and participatory research: an overview. Int. J. Action Res. **7**(2), 160–174 (2011). https://doi.org/10.1688/1861-9916_IJAR_2011_02_Thiollent

Todd, Z.: Indigenizing the anthropocene. In: Davis, H., Turpin, E. (eds.) Art in the Anthropocene: Encounters Among Aesthetics, Politics, Environments and Epistemologies, pp. 241–254. Open Humanities Press, London (2015). http://www.openhumanitiespress.org/books/titles/art-in-the-anthropocene/

United Nations. Convention on the Rights of Persons with Disabilities. United Nations Enable (2006). http://www.un.org/esa/socdev/enable/rights/convtexte.htm. Accessed Sept 2020

Wilson, L.-M.: An overview of the literature on disability and transport (2003). https://disability-studies.leeds.ac.uk/wp-content/uploads/sites/40/library/wilson-louca-DRCTransportLit review.pdf

A User-Centred Approach to User Interface Languages and Icons: Co-evaluation and Co-creation of Accessible Digital Mobility Services

Rebecca Hueting, Sabina Giorgi[✉], and Andrea Capaccioli

Deep Blue, Via Manin 53, 00185 Rome, Italy
{rebecca.hueting,sabina.giorgi,andrea.capaccioli}@dblue.it

Abstract. Challenging the acceptance of what have been defined as universal and standard pictograms, this paper promotes a conceptual approach to improve non-textual communication in digital mobility and delivery services, to ensure that different types of people may access content in an intuitive manner, overcoming language, cultural, physical and cognitive barriers. Starting from the user-centred methodological process applied in the development of the Universal Interface Language tool, one of the main outcomes of the INDIMO EU project (Inclusive Digital Mobility Solutions), this paper presents a methodological path that can provide UX/UI designers, developers and service providers with a practical guide to defining a proper set of accessible and inclusive icons as part of the user interface, be it digital or physical. In particular, this paper points out the need for bottom-up initiatives based on the co-design of physical and digital interfaces and their components to create symbols and icons with a higher degree of universality. To this end, user evaluations of mobility specific and general icons, and recommendations based on the empirical research in the INDIMO project, are presented. These address the design, selection and integration of visual icons in accessible user interfaces for digital applications.

1 Introduction

Visual icons are fundamental components of any digital application or service. They are supposed to transmit meanings in the fastest and most intuitive way. Whether we are aware of it or not, we daily interact with all kinds of icons—yet they are not accessible to all. Studies confirm that the comprehension of signs and symbols, be it in the real world or in the digital world, is not as obvious as it may seem (Bagagiolo et al. 2019).

The comprehension of icons and of the overall user interface (UI) are closely inter-related and they are both influenced by factors such as the context-of-use, socioeconomic and cultural background of users and their different levels of perception (i.e., the kind/s and degree/s of impairment). There are pictograms considered "universal", either because their visual affordance is highly intuitive for most people globally, or because they are commonly accepted as global standards and used worldwide. Yet, high degrees of affordance (Norman 1999) and acceptance are not, by themselves, sufficient

© The Author(s) 2023
I. Keseru and A. Randhahn (Eds.): *Towards User-Centric Transport in Europe 3*, LNMOB, pp. 194–212, 2023.
https://doi.org/10.1007/978-3-031-26155-8_12

to declare them universal. In this paper, the universality of the most common visual icons used by digital applications offering mobility and goods delivery services is challenged, applying a user-centred approach to their analysis and evaluation.

The purpose of this paper is to promote a conceptual approach to improve the efficiency of non-textual communication in digital mobility and delivery services (DMS and DDS, respectively) to ensure that different types of people may access content in an intuitive manner, overcoming language, cultural, physical and cognitive barriers. Starting from the methodological process applied in the development of the Universal Interface Language tool (UIL), one of the main outcomes of the INDIMO EU project (Inclusive Digital Mobility Solutions), the paper also aims at developing, refining and fostering a methodological path that can provide UX/UI designers, developers and service providers with a practical and comprehensive guide to planning, building and performing quick and intuitive exercises with users.

2   The Challenge of Using Icons, Pictograms and Signs in a Complex Society: The Universality Issue

The advantages of choosing pictograms and symbols over text to overcome the barriers of individual languages and literacy have been recognized from the beginning of the 20th century, with the *International System of Typographic Picture Education* (Isotype) method designed by Neurath. By the late 1960s, the concept of a standardised design system was considered necessary when communicating in large organisations or international events involving multilingual users (Rosa 2009). In 1972, Aichler (1996) designed a system of visual symbols "of universal intelligibility" for the Munich Olympic Games to help visitors with regard to information and communications. In 1974, the United States Department of Transportation commissioned the American Institute of Graphic Arts (AIGA) to create one of the most coherent and functional pictographic systems, in order to help large crowds easily find their way through public spaces, transport hubs and events. In the 1980s, the first international standard *ISO 7001:1980 Public information symbols,* which specified graphical symbols for the purposes of public information, was published. Even though the need for standardised and universal set of pictograms and symbols to use in international facilities or in any context involving multicultural audiences is recognized, pictograms are interpreted differently depending on culture, age, social identities and literacy levels. People's capability to read and decode graphical forms always needs to be addressed with care, since interpretation systems change across countries and cultural groups. The graphic designer Rosa pointed out how "[...] *images are always ambiguous. Like other written languages, pictograms require learning, a conscious methodology and pedagogic support"* (Rosa 2009, p. 32).

A well-known study explored the comprehension of standard healthcare symbols by a sample of participants from different cultures, age groups and literacy levels (Hashim et al.2014). The main results highlighted that: i) symbols referring to abstract concepts are the most misinterpreted; ii) interpretation rates vary across cultural backgrounds and increase with higher education and younger age; iii) pictograms with human figures and a synthetic description of actions are better understood than abstract concepts.

As argued by the Nielsen Norman Group, the universality of icons can also be questioned by considering the variability of their functionalities across digital interfaces. Most icons, in fact, "*continue to be ambiguous to users due to their association with different meanings across various interfaces*"[1].

In this regard, detailed information about the design of signs and symbols is included in the ISO 7010: 2019 (ISO 2019) safety colours and safety signs and in the ISO/IEC 11581 (ISO/IEC 2000), which provides a framework for the development and design of icons and their application on screens capable of displaying graphics as well as text.

However, the fast pace of technological innovation does not allow standards and regulations to adapt in timely fashion, and this is even truer if we look at the speed of new applications released on digital stores. The Universal Interface Language tool which we are presenting in this paper tries to bridge this gap by integrating the existing official guidance and standards with recommendations from the INDIMO research, which are useful to define a personalised step-by-step user-centred process for the inclusive design of icons and related application interfaces. As suggested by Alan Cooper, "*Obey standards unless there is a truly superior alternative*" (Cooper et al. 2014, p. 319). We argue that when it comes to visual icons, it's more likely that there is a superior alternative than that the existing standard is indeed 100% appropriate.

In many cases, users do not recognize the existing universal standard symbols as the most appropriate ones to describe a specific action, condition or need. These symbols simply do not resonate with their everyday lives and environment. The "Accessible Icon Project"[2] offers telling examples. The project started in 2010 as a "design activism" project aimed to make cities more inclusive by altering existing signs concerning the internationally adopted wheelchair-access symbol[3]. The project's founders considered this standard icon unable to represent the human body moving through space, "*like the rest of the standard isotype icons you see in the public space*" (Hendren 2016) and thus, unable to show the agency of persons with disabilities. The project team started an icon redesign process from the bottom up, directly addressing users' need to clarify a misleading representation of people with disabilities. In creating a new formal icon to replace the old one, they hired a graphic designer to bring the new icon in line with professional standards. Today, the icon is global and used in hundreds of cities and towns, at private and public organisations, and by governments and individual citizens.

This success story makes it possible to challenge the acceptance of what have been defined as universal and standard pictograms and points to the need for more bottom-up initiatives based on the co-design of physical and digital interfaces and their components (e.g., icons). Starting from users' needs and their everyday lives, such an approach could create symbols and icons with a higher degree of universality.

So, in conclusion, what about the "conscious methodology" previously mentioned by the designer Rosa? What does it consist of? In this paper, we argue that a conscious methodology needs to be based on user involvement when defining the most accessible and inclusive set of icons to include in a service digital interface.

---

[1] https://www.nngroup.com/articles/icon-usability/.

[2] https://accessibleicon.org/#read.

[3] It refers to the International Symbol of Access designed in the 1960s by Susanne Koefoed.

3   The INDIMO Project and the Universal Interface Language Tool

The Universal Interface Language (Hueting et al. 2021) is one of the main outcomes of the INDIMO project. The EU-funded Horizon 2020 project aims to extend the benefits of new digital on-demand transport or logistic services to include user groups that currently face barriers and are partly or totally excluded from using such services due to a little-inclusive design approach. Overall, the proposed project methodology consists of a user-centred approach based on the co-design of different tools with a Co-Creation Community that integrates user representatives, policy makers, academia, industry, and local Communities of Practices (CoPs) established in the project's five pilot sites (i.e., Italy, Flanders/ Belgium, Galilee, Spain, Germany).

The UIL tool has been developed throughout the three years of the project, during which the methodological steps detailed in the next paragraph (see Sect. 4), have been identified.

The UIL tool offers user-centric recommendations for the development of visual icons as part of the user interface design, be it digital or physical. It derives from the need to answer the following key questions:

- Are people aware of the emerging role of visual icons in digital mobility applications?
- How can visual icons help all people navigate smoothly within the contents and features of digital mobility applications?
- Are the meanings of visual icons clear enough to all users?

The guidelines especially address UX/UI designers, developers, and service operators. Their main purpose is to improve non-textual communication in digital applications to ensure that different types of people may access content in an intuitive manner, overcoming language, cultural, physical and cognitive barriers. The final outcome is composed of: user-centred exercises and survey templates, an icons catalogue, an icon analysis template and an extensive set of recommendations. The icons catalogue includes the involved users' evaluation of the recurring icons used in digital mobility and delivery applications. These evaluations have been collected through easy-to-deliver exercises performed in each project pilot. The systematic qualitative collection of users' evaluations of icon comprehensibility considered their perception in relation to the interface of the applications under analysis. The recommendations derived from this address the design, selection and integration of visual icons in accessible user interfaces for digital applications and organisational measures to engage users in a continuous improvement process. They also concern the lessons learnt about the user recruitment and user testing in a co-creation and co-evaluation approach, and more general tips for inclusive design.

4   Developing the "Universal Interface Language" Tool: The Methodological Approach

In the UIL, a methodological path was set up to evaluate the accessibility and inclusiveness of icons in relation to service and application interfaces. It consisted of three main steps, as follows:

1. A preliminary review of 62 digital mobility and delivery service applications (DMS/DDS) from over twenty different countries, in order to explore interface accessibility and the inclusivity of icons. In addition, another twenty applications commonly used in Europe offering both transit and food delivery services were explored. From this review, a catalogue of 27 recurring icons—both general and mobility-related ones—were identified;
2. The selection of icons collected in the previous phase was compared with those used in the pilot site applications. A UIL exercise was built to involve users in the evaluation of icons. Five similar interactive UIL exercises were performed, one for each pilot's community of practice;
3. The distribution of the online UIL survey to all stakeholders, followers and Co-Creation Community members. It aimed at bringing together the evaluations of visual icons, as collected through the previous two steps.

4.1 Preliminary Review

In the first step, a preliminary set of icons to be further evaluated in steps 2 and 3 was identified. The desk research reviewed 62 digital mobility and delivery service applications from over twenty different countries across the globe. The analysis included three main groups of applications: global routing and vehicle/ride sharing applications; digital delivery applications (including smart boxes); and local public transport (or other transport) service applications. The third group was further divided into regions, given that transport habits and regulations in different countries may vary, possibly leading to different interface designs for such applications. The testing personnel included software engineers usually working with the MBE–Budapest Association of People with Physical Disabilities, which is part of the INDIMO consortium.

The analysis mapped the existence and degree of inclusive interfaces and service solutions (e.g., public transport routes making provisions for wheelchair users), accessibility settings (personalisation accommodating specific needs), notifications (personalised info about real-time accessibility issues), voice-based options (search, route planning, navigation), also tracking personalisation options for vulnerable to exclusion groups. The analysis included the study of screenshots of the applications interfaces where both general icons and specific mobility icons were clearly identifiable. In addition, 20 applications[4] commonly used in Europe offering both transit and food delivery services were explored, along with a few applications dedicated to people with visual impairments. Based on this quite extensive list of digital applications, a catalogue of 27 recurring icons was built up (see Sect. 7.2). These were evaluated through the UIL exercises and the UIL survey, as explained in the next sections.

---

[4] The 20 common applications explored are: (DTS) blablacar, Cabify, Citymapper, Flixbus, FreeNow, Lyft, Moovit, Omio, Safr, Transit, Uber, and Waze; (DDS) Deliveroo, JustEat, Glovo, and UberEats; the apps for the visually impaired are BeMyEyes, Emit, Kimap, and Wheelmate.

5 The UIL Exercise in the INDIMO Pilot Sites: A Co-evaluation Experience

The selection of icons collected in the first step was compared with those used in the pilot sites' digital services. The pilot services consist of: a digital locker to enable e-commerce in rural areas (Emilia Romagna - Italy); inclusive traffic lights (Antwerp/Flanders); informal ride-sharing in multicultural towns (Galilee); cycle logistics platform for food delivery (Madrid - Spain); on-demand ride-sharing integrated into multimodal route planning (Berlin-Germany).

Thereafter, five similar interactive UIL exercises were built and performed starting from existing Human Factors design and UX testing examples (Bagagiolo et al. 2019) (ETSI 2002–2008), and questionnaires (Blees and Mak 2012) (Zender and Cassedy 2014). The UIL exercises took place in the pilots' communities of practice between March and April 2021. A total of 46 participants attended, including: users and non-users, civil society organisations representing groups vulnerable to exclusion, operators, policymakers, researchers, and other relevant stakeholders. Six recurring icons were tested in the pilot applications, and the participants of local communities of practice cooperated in evaluating the icons' use within the application interfaces during 1-h online sessions. An interactive Miro dashboard supported the activities being performed; in two cases, to accommodate specific users' needs, one-to-one sessions were organized to mitigate the "technological barrier" of the online board.

The participants in the pilots were asked to share their experience and comment about: i) the meaning of the icons; ii) potential matching with other icons that could be used to convey the same meaning; iii) elements that were unclear or produced confusion in the visual outlook; iv) elements that could be added for clarification or sharper communication; v) other elements that should be kept in consideration when designing graphic interfaces.

To ensure the highest degree of inclusivity during the meetings—which were held online, due to the pandemic situation—the interactive exercises were led by a guiding moderator who presented the slides on screen and facilitated an open discussion verbally.

The exercise consisted in two parts: one introducing the theme of the ambiguity of icons and the other exploring their use in the digital context of the application itself.

The first part consisted of the "icons pitch". All the participants were shown a first set of icons that are typically part of the graphic language of most mobile apps and a second set of matching icons with similar meanings. Figure 1 shows a screenshot of the icons pitch. It includes the general icons to be tested, and the main points discussed (e.g., pictograms' meanings according to the participants' experience, matching with other icons with the same functionality, conflicting meanings, etc.).

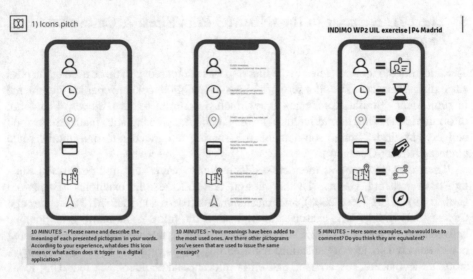

Fig. 1. An example of UIL exercise – icons pitch

In the second part, participants were invited to observe the same icons as they appeared in the different pilots' application screens and to comment on them. Figure 2 shows the use of the pinpoint icon in the Berlin pilot application.

The exercises provided a clear understanding of the common interpretations that people give to visual icons, the variety of meanings attached to them, the interaction between their intrinsic characteristics and the relationship with the other components of the user interface.

Fig. 2. An example of the UIL exercise - Berlin app screens

6 The Online Survey Involving Stakeholders and the Co-creation Community

The third and final step was that of collecting more quantitative data about visual icons through an online survey. This aimed at exploring end users' comprehension of the visual icons most used in digital mobility and delivery services. The UIL survey complemented the icon-evaluation results collected through the review of DMS and DDS applications, and the UIL exercises performed in the local CoPs. The survey was distributed online to all project partners and stakeholders, social media followers, and to the members of the INDIMO Co-Creation Community. Different sections were included to investigate: i) the perceived comprehensibility and ambiguity of recurrent pictograms in digital mobility and goods delivery services (see Fig. 3 and 4), ii) the overall perception of such services' accessibility, and iii) respondents' socio-economic background (e.g., age, gender, education, employment status, caregiving activities, income). The collection of responses lasted for three weeks in May 2021. In total, 89 responses were gathered. A frequency analysis was performed. The full set of results is included in the INDIMO project deliverable "Universal Interface Language – Version 1" (Hueting et al. 2021).

Sample questions related to the first section of the survey are reported in Fig. 3 and 4.

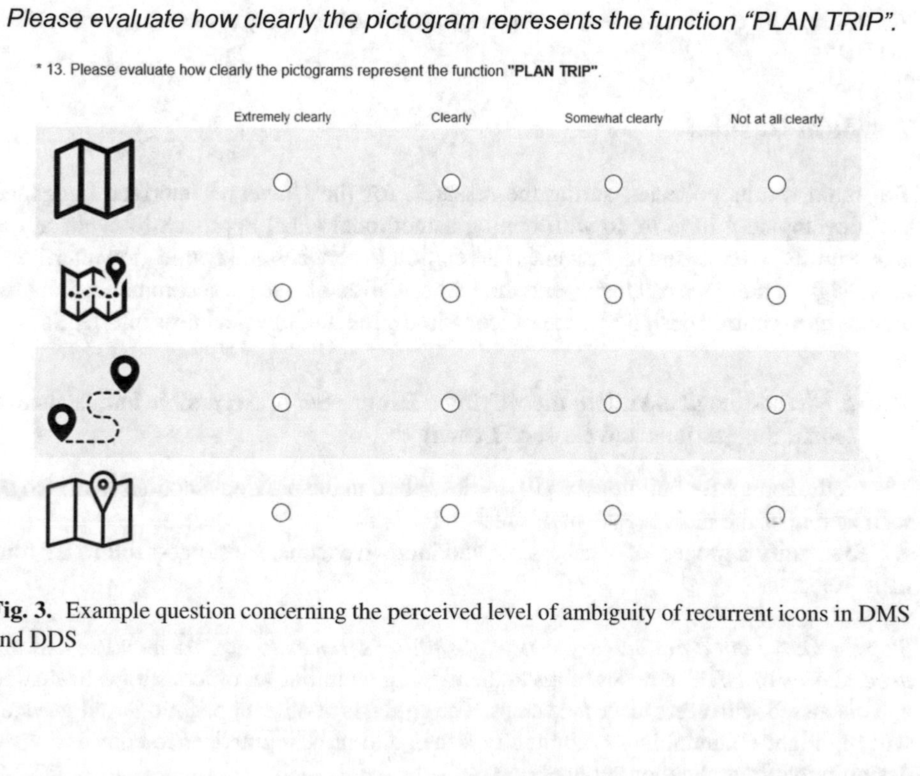
Please evaluate how clearly the pictogram represents the function "PLAN TRIP".
* 13. Please evaluate how clearly the pictograms represent the function "PLAN TRIP".
Extremely clearly
Clearly
Somewhat clearly
Not at all clearly

Fig. 3. Example question concerning the perceived level of ambiguity of recurrent icons in DMS and DDS

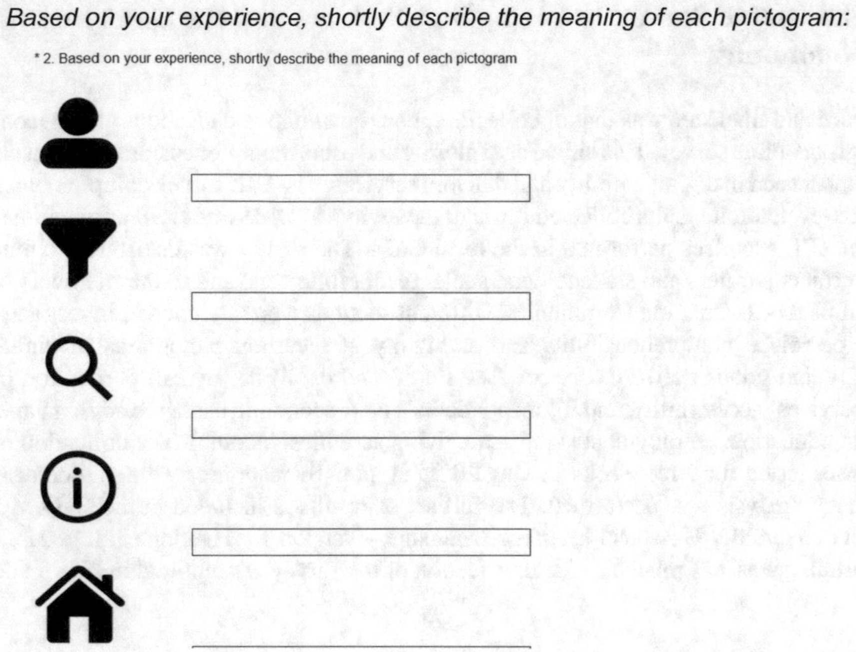

Based on your experience, shortly describe the meaning of each pictogram:

* 2. Based on your experience, shortly describe the meaning of each pictogram

Fig. 4. Example question concerning the perceived comprehensibility of recurrent icons in DMS and DDS

7 Main Results

The main results collected during the research for the Universal Interface Language tool development have to do with honing a methodological approach to evaluate the accessibility of icons and inclusiveness in relation to services and application interfaces, the design of an icons catalogue, and the identification of a set of recommendations for a more user-centred design and use of icons in digital and physical user interfaces.

7.1 A Methodological Path to Identifying a Proper Set of Accessible and Inclusive Icons: Suggestions and Lessons Learnt

The methodology for building the UIL as described in the previous sections, can also be seen as one of the main results of our research.

To identify a proper set of accessible and inclusive icons, we suggest following four main steps.

Step 1 - Carry out a preliminary review of similar services to explore icon use and the accessibility of UI. The review aims at identifying an initial set of icons to be evaluated and discussed with users in the next steps. The analysis of other applications and services will highlight potential icons' ambiguity issues and/or best practices to consider when designing your application/service.

Although the review does require considerable time and effort, this first step is crucial for identifying an initial set of icons to work with, and becoming aware of icons' ambiguity across different digital interfaces. Furthermore, no existing reviews or literature are available in this regard, except for some icon usability studies[5,6] published online by the UX designer community.

Step 2 - Build a user-centred exercise to involve users in the co-evaluation of icons related to UI. Perform a quick and simple exercise with users to assess the comprehensibility of icons in relation to the application interface. The exercise focuses on collecting feedback about the user experience, from the viewpoint of vulnerable users. The main objectives are the following:

- Raising participants' awareness (of both users and developers) concerning the ambiguity of icons;
- Identifying the most common issues in the usability of icons;
- Identifying how the application interface and internal structure influences the comprehension of icons;
- Finding potential solutions or mitigations to accessibility barriers in digital applications.

Participants will provide you with feedback about:

- The meaning of the icons;
- The potential matching with other icons that could be used to convey the same meaning;
- Elements that are unclear or create confusion in the application interface;
- Elements that could be added for clarification or more accurate communication;
- Other elements that should be kept in consideration when designing a user interface.

Step 3 – Consolidate the results of the preliminary review and exercises with a UIL survey. A survey can be a useful tool to complement the icon-evaluation results collected in the exercise. Such a survey explores the use of pictograms in digital mobility and goods delivery services mainly by using four-step Likert scales, thereby identifying which icon best represents the given function with the least ambiguity (see Fig. 3 as example). The survey should be answered by all users, especially involving those categories who experience some kind of barrier in using such an application/service (people with different kinds and degrees of impairments).

Step 4 - Organise the results into an icon catalogue to aid the improvement of the chosen set of icons. The final outcome of this suggested methodology should be an in-depth evaluation of the identified icons. This can be organised into an icon catalogue integrating the main results collected throughout the different steps. The catalogue can

---

[5] https://www.usertesting.com/blog/user-friendly-ui-icons - last access on 29th June 2021.

[6] http://babich.biz/icons-as-part-of-an-awesome-user-experience/ - last access on 29th June 2021.

be further improved and updated with time, and used as internal reference for all further services which the designers or developers team want to analyse or enhance. The next section will describe how the icon catalogue was built up in the INDIMO research.

Especially in steps 3 and 4, the involvement of users and other significant stakeholders is crucial for successfully achieving the aim to co-create and identify the most appropriate set of icons for a service or application. From the INDIMO experience, we learnt some key lessons in this regard.

First of all, involving vulnerable-to-exclusion users requires a fair amount of time and sincere commitment since it is not easy to identify and convince them or their representatives to participate.

Recruitment and relationship building could focus more on creating an open and direct relationship with the people involved, both end users and their representatives, to ensure continuity and trust building.

Due to the COVID-19 pandemic, it has been a challenge to engage people with the characteristics that we wanted to address in online meetings and activities. Many people declared connection fatigue. A small positive remark is that, for some people, it was easier participating directly from their homes instead of being required to come to a meeting in person.

Finally, tools for online events are not yet accessible enough and there are few and poorly designed alternatives overcoming such barriers (live captions, interactive boards, WebEx meeting platforms). In many cases, WhatsApp and Zoom turned out to be the most feasible options, since they were already used on a daily basis by most users.

Despite the above considerations, the project consortium members, the authors of this chapter and all involved participants were surprised by the richness of insights gained from the discussion about the applications' usability and the icons used during the UIL exercise sessions. An open and non-judgmental setting was an important feature, together with the preliminary knowledge participants had about the applications. These allowed more problems and consequent ideas for improvement to emerge.

7.2 The INDIMO Icon Catalogue

As a result of the extensive desk research across the digital mobility and goods delivery applications, the exercises performed in the pilots' communities of practices, and the online survey distributed among the INDIMO consortium and the Co-Creation Community, a total of twenty-seven common icons were identified. Figure 5 shows the 27 identified icons. It includes the pictograms and their related functionalities (e.g., timetable). The catalogue includes both mobility specific icons and general icons. For each icon, the catalogue reports: a) examples and participants' evaluations of the icons used in the pilot applications and collected during the UIL exercises; b) a summary of the results from the UIL survey concerning the icons' comprehensibility.

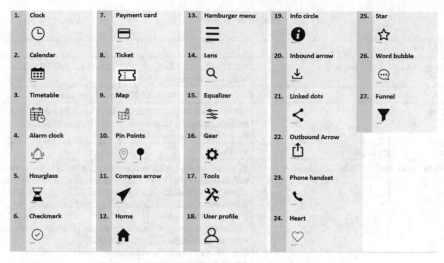

Fig. 5. The INDIMO icon catalogue

The following table reports two examples from the INDIMO icon catalogue. The first concerns the "clock" as a specific mobility icon, the second the "tools" as a general icon[7].

Table 1. Examples fom INDIMO Icon catalogue with pilot applications

Icons' group, name and function	1. Clock
UIL Exercises results: Galilee, Madrid, Berlin pilots	In transport apps, this icon may represent schedules or calendars, and maybe also the availability of service/opening hours, and the expected time of arrival of the ride-sharing vehicle. In delivery apps, it could represent the waiting time before preparation, the time of delivery or the opening hours of the restaurant/food provider It is not clear if the clock is associated with the departure or arrival time. The fact that it is used in multiple ways creates confusion. Unfortunately, there is no text label in most screens where it is used. It would be clearer with numbered hours UIL survey results – Q14 - Please match the pictogram with the function it best represents, based on your personal experience - 56% (of responses) - "Set alarms"; - 20% - "View current timing" of something/ someone arriving or of an item that has to be delivered; - 10% - "View expected date /time of arrival/ delivery"

(continued)

---

7 The full content of the catalogue is included in the "Universal Interface Language – Version 1" (Hueting et al. 2021).

Table 1. (*continued*)

Icons' group, name and function	1. Clock
Example of use (Madrid pilot):	"La Pajara" food delivery app order screen using the clock icon to notify time of delivery (see Fig. 6)
Pros:	The contrast and position of the icon is appropriate
Cons:	The size is too small for users with low vision; there is no textual explanation and the same icon is used with multiple meanings in the same app, affecting consistency
Icons' group, name and function:	17. Tools
UIL Exercises results: Emilia-Romagna pilot	This icon is clear, but it is not clear who is going to fix the problem with those tools. To low-digital-skilled users it is ambiguous since it does not tell them if it is there to offer external assistance or if she/ he (the user) should operate with (digital) tools and try to solve problems (see Fig. 7) UIL survey results Q4 - Please evaluate how clearly the pictograms represent the function "GO TO SETTINGS" The tools icon is considered the second choice for access to settings functions (68% of positive preferences), after the gear icon Example of use (Emilia-Romagna pilot): "Punto Poste da Te" (digital locker service) app error screen using the tools icon to notify the user that the process has been interrupted
Pros:	The design of the tools is clear; contrast and size are appropriate and the textual description supports comprehension. The spacing across elements is appropriate for all users, including those with digitisation issues. The icon associated with the cross in a red circle helps the user understand that something is wrong
Cons:	The icon meaning cannot be read by text-to-speech systems. Since this function is related to parcel delivery in mailboxes, users with low digital skills or non-native speakers may think that someone will physically provide help to adjust/recover the service

7.3   Recommendations for the Design of Accessible and Inclusive Interfaces

Recommendations based on the empirical research in the INDIMO project address the design, selection and integration of pictographic icons in accessible user interfaces for mobile applications. They also include those derived from the literature, standards, desk research and tips for user-testing and recruitment.

Since the main reference for web-developers and designers in terms of web accessibility are the WCAG guidelines, we chose to use the same categorisation to offer UIL readers a sample of recommendations aligned with its structure: perceivable, adaptable, robust, operable and understandable interfaces (W3.org 2018).

From the INDIMO user research the following accessibility topics emerged as the most problematic, and they have been grouped as follows:

Perceivable Interfaces

- *Pay attention when designing welcome screens*: Welcome screens are often overlooked, despite being the first hook for catching users' attention. The service provided should be quickly recognised, clearly stated and the navigation facilitated by labels and tips. Especially in the first screens, it is important to avoid information overload.

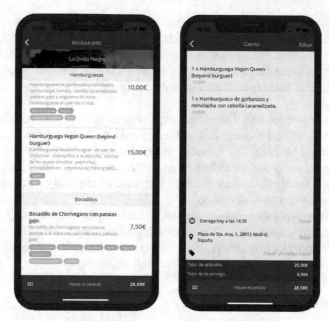

Fig. 6. Example of use for the tool "clock" in Madrid pilot

Fig. 7. Example of use for the tool "tools" in Emilia-Romagna pilot

Thus, it is important to provide direct access to the few features needed to easily access the service.

- *Colour coding*: Colour coding brings a lot of information to users, though this is very much related to culture. Global standards and guidelines fail in providing information on this concern. The use, misuse and non-use of colour can be misleading in different ways, depending on the context of use and socio-cultural environment. Test colour palettes of buttons and icons with diverse people with different socio-cultural backgrounds.
- *Themes and backgrounds*: Situational impairments or changing conditions should be considered when choosing colour themes and backgrounds of user interfaces. Solutions could be identified for other contextual conditions affecting perception, for all kinds of users and on different sensory channels (e.g., road navigation maps offering light-sensitive backgrounds, which change dynamically when sensors detect low-light conditions such as tunnels).
- *Contrast*: contrast is very important to colour-blind people: to avoid losing the information conveyed by colours, and misunderstandings, developers should design mock-ups in grayscale and choose colour palettes and opacity options based on the preliminary results of online contrast accessibility checking tools.

Adaptable Interfaces

- *Sorting*: pay attention to lists/elements sorting: they have to make sense depending on use frequency and process priority, not alphabetical order; whenever possible, offer the option to choose sorting order and filter options.
- *Accessibility settings*: offer the possibility of partly customise the theme (colour, contrast, fonts and content behaviour), provided that at least one of them is fully-accessible.
- *Inclusive fields/travel options:* when asking users to fill in information about themselves or about service requirements/ preferences, add inclusive fields where users can specify additional needs.

Robust Interfaces

- Constraints: add constraints to ensure that users are not required to insert identical data multiple times, offering options to verify, edit or specify changes before proceeding with transport or delivery service order confirmation.
- Provide automated error detection function: provide help buttons, show how the error detection works, explore advanced strategies for error-prevention, such as user input constraints (e.g., poka-yoke).
- User rating and reviews: offer users the possibility to rate and review your application in terms of accessibility, with a transparent and open app-rating area, and be sure to address/solve the issues that may emerge. Consider this an opportunity for further improvement.
- Long-term user engagement strategies: think in a long-term perspective and invest a proper amount of time, money and effort to testing ideas and prototypes, taking advantage of the experience and knowledge of the real experts—that is, those users experiencing accessibility barriers who can provide real feedback about services.

Operable Interfaces

- Increase broad operability: ensure that your service, both in its desktop and mobile versions, is operable and compatible with as many devices, operating systems and browsers as possible and that it is easily accessible also in different contexts of use. Develop a light version of the App which is operable with a variety of equipment, including older models of devices, and which does not require too much storage space or operating memory (RAM) nor overly affect battery consumption. A web-based alternative should be available.
- Transparency about accessibility limits: be honest about your limits and offer users explanations concerning the accessibility limits of your services, either through the frequently asked questions (FAQ) section or a dedicated area for open comments or specific complaint forms.

Understandable Interfaces

- *Easy-to-read privacy policies and terms of use*: full and easy access to your Privacy Policy, Terms of Use and Personal Data Treatment information should be provided to all users, mitigating readability issues through easy-to-read texts, visual explanations and simplified navigation across contents (text blocks, sections).
- *Quick and easy editing of personal data treatment settings*: personal data should always be available for users to edit, posing no time-limits and no risk of data loss during compilation. Also, a higher level of control and support should be available, for example by offering direct links to organisations that users can call anonymously to receive help (trusted referees).
- *Tips*: At key steps, ask the user for suggestions or complaints concerning the service; send e-mails to users to ask for their ratings and comments about service quality and satisfaction, allowing in-message reply; include the possibility of viewing other users' rating of the service; include a service agent and offer users the possibility to directly contribute to the FAQ area.
- *Tutorials*: realise first-use tutorials in different media formats, languages and easy-to-read textual contents, to ensure all users find the ones most appropriate to their needs (textual, audio, images, hard printed/printable copies), including the option to skip.

Concerning our focus on inclusive visual icons, recommendations include results derived from the Communities of Practice meetings in pilot sites and the UIL survey, i.e.:

- Choose icons describing actions rather than objects or symbols;
- Ensure the internal consistency of the icons;
- Label icons;
- Get familiar with naturalistic observation of users interacting with icons;
- Test iteratively for recognisability and memorability;
- Consider a flat and minimalistic design of icons;
- Keep in mind that ambiguity may increase over time or be misunderstood in diverse socio-cultural context;

- Use skeuomorphism only if it is essential;
- Limit the use of animated icons.

Last but not least, we shared our lessons learnt in terms of user recruitment and testing in the service prototyping or co-design phases. This sub-set of recommendations suggests that it is important to:

- *Build a network of people* who will participate in design iteration through co-creation workshops and interviews from the early phases. Such collaborative communities are intended to ensure cooperation across developers and design experts and all the potential customers left out by traditional user-testing, namely the people who experience barriers in using digital mobility applications. Due to these barriers, they are poorly reached by market-oriented campaigns, so a targeted strategy should be put in place.
- *Reach out to local advocacy group*s in advance, as they may help get in contact with users willing to share their experiences and as a result have their voices heard in the development process, in order to build inclusive-by-design services.
- *Organise both online and real-life meetings and workshops* to tackle potential barriers to inclusion; if you really want your participants to enjoy the activities you organise and provide honest feedback, you should share and verify workplans in advance and collect their suggestions prior to the meeting. Empathise with involved people's needs and you'll be rewarded by the experience.
- *Define a clear, simple and continuous inclusive design process*, using existing templates or building your custom set. Instruct team members to follow the same guidelines and track results in the most efficient and systematic way possible.

This initial set of recommendations grew considerably during the research. It has been collected as a browsable online catalogue that integrates all the recommendations derived from the INDIMO research[8]. The main ambition of the INDIMO set of recommendations is that of guiding UX/UI designers, service providers and developers in a truly inclusive approach to the design of digital mobility services.

8 Conclusions

When people or objects travel, they move across the four dimensions of space and time. How can icons representing objects or actions related with time and movement be designed in such a way that they are clear enough and unambiguous for all people over time and across countries? The fascinating history of signs and symbols is a never-ending one. The challenge remains open, and the main lesson is that no universal icon can be defined once and for all.

Nowadays, thanks to digital technologies, we are finally able to create dynamic contents and adapt all services and related applications to the changing needs of all kinds of people. When referring to needs, it is essential to include those expressed by the 15% of world population living with some kind of impairment. Undoubtedly, the usability of

---

[8] To see the full set of the INDIMO recommendations, see the online catalogue at https://spet. indimoproject.eu/recommendations/.

every icon and user interface can be increased by applying user-centred design techniques and the Universal Interface Language approach proposed by the INDIMO project. But without the direct involvement of the targeted end users, they will have a limited reach, since without such a commitment to involving them, even good ideas will surely be poorly implemented. Thanks to co-creation, a direct exchange between end-users and service design teams was enabled, allowing the people involved to become familiar with each other and grow their awareness about their mutual needs and the complexity of the job at hand. This ultimately activated empathy, awareness and a commitment for change and adaptation. The main lesson learnt—as reported by the service operators, designers and developers involved as members of the INDIMO pilots' communities of practice in the co-evaluation and co-design process—is that having the opportunity to learn directly from these usually neglected end users brings a win-win situation. For the end users gain more satisfaction from using digital applications, and service providers, developers and designers improve the quality of their service, thus reach better market positioning while also growing personally. The INDIMO Universal Interface Language tool's main ambition is exactly that of providing practical tools to bridge the "communication" gap between accessibility needs and the diverse solutions in the digital world. The INDIMO results can represent the starting point for further research and developments, and help others unleash the huge potential of collaborative design-for-all.

References

Aichler, O.: Zeichensysteme der visuellen Kommunikation - Handbuch für Designer, Architekten, Planer, organisatoren. Ernst & Sohn (1996)

Bagagiolo, G., Vigoroso, L., Caffaro, F., Micheletti Cremasco, M., Cavallo, E.: Conveying safety messages on agricultural machinery: the comprehension of safety pictorials in a group of migrant farmworkers in Italy. Int. J. Environ. Res. Public Health **16**(21), 4180 (2019). https://doi.org/10.3390/ijerph16214180

Blees, G., Mak, W.: Comprehension of disaster pictorials across cultures. J. Multilingual Multicultural Develop. **33**(7), 699–716 (2012). https://doi.org/10.1080/01434632.2012.715798

Cooper, A., Reimann, R., Cronin, D.: About Face 3: The Essentials of Interaction Design. Wiley Publishing Inc., New York (2014)

European Telecommunications Standards Institute (ETSI) (2002–2008). Human Factors (HF); Guidelines on the multimodality of icons, symbols and pictograms (No. ETSI EG 202 048 V1.1.1). https://www.etsi.org/deliver/etsi_eg/202000_202099/202048/01.01.01_60/eg_202048v010101p.pdf. Accessed 5 May 2021

Hashim, M.J.: Interpretation of way-finding healthcare symbols by a multicultural population: navigation signage design for global health. Appl. Ergonomics **45**(3), 503–509 (2014). https://doi.org/10.1016/j.apergo.2013.07.002

Hendren, S.: An icon is a verb. About the project (2016). https://accessibleicon.org/#an-icon-is-a-verb. Accessed 7 April 2022

Hueting, R., Giorgi, S., Capaccioli, A., Bánfi, M., Soltész, T.D.: Universal Interface Language – Version 1. INDIMO project deliverable (2021). https://www.indimoproject.eu/wp-content/uploads/2022/02/INDIMO-D2.3-FINAL_v2.0.pdf

ISO - International Standard Organisation (2019). ISO 7010:2019 - Graphical symbols — Safety colours and safety signs — Registered safety signs. https://committee.iso.org/standard/72424.html. Accessed 4 Feb 2021

ISO/IEC (2000). ISO/IEC 11581–1:2000 - Information technology — User system interfaces and symbols — Icon symbols and functions. Part 1: Icons – General. https://www.iso.org/standard/24267.html. Accessed 4 Feb 2021

Norman, D.A.: Affordance, conventions, and design. Interactions **6**, 38–43 (1999). https://doi.org/10.1145/301153.301168

Rosa, C.: 40 years of Pictograms in universal contexts. What's next? (2009) https://doi.org/10.13140/2.1.2205.6967

W3.org. Web Content Accessibility Guidelines (WCAG) 2.1 (2018). https://www.w3.org/TR/WCAG/. Accessed 12 May 2021

Zender, M., Cassedy, A.: A (mis)understanding: Icon comprehension in different cultural contexts. Visible Lang. **48**(1), 68–95 (2014)

Decision Support for Policy Makers for Inclusive and Accessible Mobility Services

Subjectification, Technology, and Rationality – Sustainable Transformation of the Mobility Sector from a Governmentality Perspective

Julia Hansel[(✉)] and Antonia Graf

Institute of Political Science, University of Münster, Münster, Germany
julia.hansel@uni-muenster.de

Abstract. Shared mobility services play an essential role in a sustainable mobility transition and unfold among so-called smart technologies. Although this can positively affect mobility, it also poses challenges for the development of sustainable urban mobility, for example, because the smart options are not equally available to all people or are inaccessible. Issues of social or ecological inequality as well as the digital exclusion of people in the mobility sector are increasingly becoming the focus of attention. Largely unexplored in this context is how the subjects of shared mobility services will be conceived, and what knowledge, skills, and resources they should bring to use smart and shared mobility services in the future. We contribute to closing this research gap by investigating the rationalities that sustainable smart and shared mobility transformation follow, which developments are triggered by the technologies, and in which ways identification offers address subjects. Foucault's concept of governmentality is used as a theoretical perspective and nuanced with critical (feminist) literature on identity formation. Methodologically, this article works with qualitative content analysis of policy documents and an ethnographically oriented observation of registration conditions in various car-, bike-, electronic moped, and scooter-sharing services. The results show that subjects are addressed in a rather general way, and their (special) needs are hardly considered. Instead, they are addressed as flexible citizen-consumers and correspond with the rationality of (green) economic growth and the liberal paradigm. Accordingly, the technologies aim for innovation, fair competition, and the provision of public space by the state.

1 Introduction: Sustainable and Smart Mobility Policy[1]

With the Paris Agreement on Climate Change (2015) and the Sustainable Development Goals (SDGs), fundamental aims for climate protection and sustainable development were agreed upon internationally. In the transport sector, greenhouse gas emissions

---

[1] We thank the reviewer for the relevant comments. The mobility plans analyzed here, are subject of another article in German with a similar question. The article is currently under review.

© The Author(s) 2023
I. Keseru and A. Randhahn (Eds.): *Towards User-Centric Transport in Europe 3*, LNMOB, pp. 215–234, 2023.
https://doi.org/10.1007/978-3-031-26155-8_13

are caused mainly by private motorized transport, which (still) continue to increase. In addition, there are negative externalities due to sealing, noise, air pollution, and land consumption. In urban areas, in particular, private motorized transport places an additional burden on the already scarce resource of available urban space (European Commission 2019).

The transformation of the mobility sector is an essential part of a (strongly) sustainable, i.e., ecologically sensible and socially just, development in urban areas and is addressed with increasing urgency as a political task (Banister 2008; May 2013). For example, the EU Clean Energy for all Europeans Package (EU 2018/1999) stipulates the preparation of National Energy and Climate Plans (NECPs). In these plans, all member states commit to reducing greenhouse gas emissions. Germany aims to reduce emissions in the transport sector by 40–42% by 2030 (German Federal Government 2019). Numerous cities in Europe and many German municipalities are looking for environmentally sound solutions for updating urban development and mobility plans. It is increasingly discussed to switch from private cars to extended environmental modes (public transport, cycling, and walking as well as shared forms of mobility) under the term multimodality. In this context, shared mobility services offer higher flexibility and enable users to switch smoothly between different modes according to individual travel needs.

The political endeavor for multimodal sustainable mobility options is often realized with internet-based, i.e., digitally supported (smart) solutions. *The Smart and Sustainable Mobility Strategy* of the European Commission (2020) proposes a "twin transition" of sustainability and digitalization to reshape and economically revitalize the transport sector[2]. This includes new mobile services[3], particularly the so-called shared (micro-) mobility (Docherty et al. 2018). Sharing options, such as car- and bike-sharing systems as well as electronic scooters and mopeds-, promise to be easily accessible and ecologically more sensible alternatives to private cars. These options are accessible to users via GPS-supported location and digital booking and billing systems. In this way, sustainability strategies for shared modes of transport are closely linked to the policy field of digitalization.

Approaches such as the new mobilities paradigm or mobility justice understand mobility as the potential for movement (motility) that is not necessarily limited to a physical change of location; communication is viewed as a journey of information, and speechlessness is understood as immobility (Sheller and Urry 2006, 2016). The ability to move of different social classes, gender, or ethnic affiliations thus becomes just as much a focus of attention as the connection between socio-demographic factors (such as income, education, and health) and (im)mobility (Lucas 2012; Sheller 2018, Martinez and Keserü in this volume). Forms of discrimination and disadvantage are thus inscribed in people's mobility behavior and their access to mobility options and are regarded as a power-laden and political phenomenon (Cresswell 2010).

---

[2] "The transition to safe, accessible, inclusive, smart, resilient and zero-emission urban mobility requires a clear focus on active, collective and shared mobility underpinned by low- and zero-emission solutions" European Commission (2021, 2f.).

[3] Mobile services are discussed under the catchword Mobility as a Service (MaaS), including traditional businesses such as taxis, but also new business areas of the sharing economy, leasing models or future services such as travel with autonomous shuttles.

Using Foucault's (2004a, 2004b) concept of governmentality and Judith Butler's (1995) work on identity formation, we follow the few examples of constructivist mobility research. In this paper, the digitally supported sustainable transformation of the mobility sector is consequently understood as a socio-cultural and political negotiation process, including the constitution and reproduction of identity options (Deffner; Sonnberger and Graf 2021). While technology is often in focus, potential users of smart and shared transport options are mostly left out in the political and scientific debates. Consequently, it is mainly unexplored which knowledge, qualifications, and resources users of shared mobility services should and must bring actually to use them. In other words: how are they conceived as subjects of new technologies? Moreover, considering transport development, it usually remains open to which rationalities the transformation follows and how technologies unfold in society. We address this research gap by investigating who smart and shared mobility services target and with what intention (rationality). Also, we examine what preconditions are tied to the use of technologies and what consequences this has for the formation of subjects. Therefore, we ask: Which rationalities does the sustainable smart mobility transformation follow? Which developments are triggered by technologies? And how are subjects addressed and identity options offered? This article aims to make these processes associated with the transformation of mobility visible, question them in terms of their steering effects, and ultimately outline them regarding the goal of a sustainable and inclusive mobility transformation. Local mobility plans of three German cities and ethnographic observation of registration requirements for different sharing services are used for an empirical illustration in this paper.

The article is structured as follows: first, the background to shared mobility is explained and critically classified in section two. Then, the theoretical framework based on the concept of governmentality and the applied heuristics is explained. The methodological approach is described in the fourth section. In the fifth section, the analytical results are presented in three subsections. Finally, we discuss our results concerning the research question in the conclusion.

2 Mobility, Transformation, and Shared Mobility

2.1 Mobility Instead of (Only) Traffic

Transport is considered as the physical manifestation of mobility needs and, thus, the actual realization of ways (Mattioli 2016). Mobility instead includes a comprehensive system of socio-cultural, technical, political, and legal factors, which together result in motility, which is the potential to be mobile. Mobility is a means of achieving and fulfilling everyday actions and needs (Mullen and Marsden 2016). Consequently, social science research is increasingly looking at mobility as a sociocultural system linked to *agency* (Graf and Sonnberger 2020; Sonnberger and Graf 2021).

The new mobilities paradigm (Sheller and Urry 2006, 2016) views social and physical processes of mobility as embedded in social structures. In this understanding, Cresswell (2010) addresses the political dimension of mobility. Like other domains, mobility is influenced by social factors and relations such as class, gender, ethnicity, religious affiliation, etc. Accordingly, mobility can be understood as a resource accessed and perceived differently by different actors. Hence, people have different kinds of access to mobility as well as experiences with it.

Direct and indirect mobility-related disadvantages are discussed under the term *transport poverty*. Lucas (2012) describes how social factors can link to more difficult access to mobility options and thus lead to mobility poverty (transport poverty), which in turn leads to immobility and thus - in a circular fashion – again to inaccessibility of places and services. These forms of exclusion often interact with other categories of difference and mutually reinforce one another (Lucas 2012). Therefore, mobility behavior can also be reflected in affiliations, language barriers, implicit codes of conduct, social networks, or value systems (Priya Uteng 2009). Additionally, there are fear-based as well as physical and space-based exclusions, which in turn can show unequal access to mobility options and thus can ultimately lead to different forms of immobility (Médard de Chardon 2019; Lubitow et al. 2020; SHARE-North 2021). The usage of shared mobility services depends on the digital skill of people, which can lead to further disadvantages in terms of these modes (Groth 2019; Horjus et al. 2022).

2.2 Transformation of the Mobility Sector

As described above, multimodality intends to facilitate the switch from environmentally harmful and space-consuming car use to more sustainable mobility options. This means that journeys should not be made with one vehicle alone but with multiple different modes of transport instead. For example, a car journey is replaced by walking to the next public transport stop from where a train or bus takes subjects to another stop, where they can reach their actual destination with a bicycle - possibly a shared bike[4]. In this context, sharing services in the field of (active) mobility play a significant role. They can be used for the flexible realization of the so-called 'first/last mile' to get from a train or bus station to the destination (European Environment Agency 2019). Proponents of a smart transition of the mobility sector describe:

"a vision of the future in which mobility will be framed as a personalized 'service' available 'on demand', with individuals having instant access to a seamless system of clean, green, efficient and flexible transport to meet all of their needs" (Docherty et al., pp. 114f.).

Along with socio-technical transitions towards smart mobility come changes in the governance of such systems. Marsden and Reardon (2018) describe the changing role of state power so that different spatial and functional networks of public, private, and non-governmental organizations come into focus of the analysis. Secondly, a change 'from ownership to usership' is described. Consequently, the marketplace of mobility services is also changing fundamentally. Individual travel and travel times are becoming increasingly commoditized, which could further fuel the long-term trend of neo-liberalization of the mobility sector (Gössling and Cohen 2014).

---

[4] Shared means of transport, such as the rental bike at the train station or the car in car sharing, rely on the principle of use without linking this to ownership at the same time. They thus touch on the area of the sharing economy and, depending on their orientation, are located at different points on the continuum between non-commercial, partly solidarity-based and partly dissident initiatives and commercially constituted services, partly belonging to large companies.

Besides digitalization dynamics, the mobility sector is also being transformed by laws such as the German Car-Sharing Law (CsgG) or the Electric Mini Vehicles regulation (eKFV), which regulates the introduction of small electric vehicles (such as electric scooters) at the federal level in Germany. An amendment of the Passenger Transportation Law (Personenbeförderungsgesetz – PBefG) of 2013 sets the goal of complete accessibility in public transport by 2022. It thus integrates the inclusion of people with disabilities into binding federal legislation. So far, citizens are perceived as users and as a source of mobility data that is collected automatically (Docherty et al. 2018). Shared mobility should enable smart as well as ecologically and socially sustainable mobility options. In the long term, a kind of networked ecosystem[5] of different mobility services could emerge in which the boundaries between various forms of mobility seem to merge fluidly into one another (Hietanen 2014). These developments result in the greater significance of shared mobility services.

2.3 Critical Reflections on Smart and Shared Mobility

Following critical research on ecological modernization and its linkage to the logic of (neo)liberal market economies (Hajer 1997; Schwanen et al. 2011), the proximity to market-based and technology-based solutions is also problematized concerning smart and shared mobility. Gössling and Cohen (2014) describe an optimism toward technological innovation that is, at least in part, the product of strong interest groups. In this context, state actors are primarily assigned the role of facilitating innovation and creating market-based regulatory approaches. On the other hand, behavioral changes should be chosen as voluntarily as possible by subjects. Politics on smart mobility tends to emphasize the role of consumers as end-users of a service, so-called citizen-consumers (Mattioli and Heinen 2020). This might result in a stronger focus on user-friendliness than democratic values (Kronsell and Mukhtar-Landgren 2020). In the context of MaaS, Pangbourne et al. (2020) describe potential ideological pressure toward governance to enable revenue streams out of previously public goods, such as public space. This could result in increased neglect of social and ecological sustainability.

Smart mobility is often envisaged as a solution that enables mobility and carbon emission reductions because mobility is expected to be electrified, shared, and more efficient. Following this logic, achieving smart mobility is often expressed as a goal on its own (Paulsson and Hedegaard Sørensen 2020). Still, it is argued that smart mobility can fulfill its desired societal outputs if steered in that direction (Docherty et al. 2018). Reliable measures towards smart and shared mobility must be actively brought in line with the sustainability paradigm instead of following the logic of an automatic equation (Lyons 2018; Heinen and Mattioli 2019; Paulsson and Hedegaard Sørensen 2020).

3 Governmentality as a Perspective on Sustainable Mobility Transformation

Constructivist or post-structuralist approaches have found their way into mobility research but are still rare. Although there are some exceptions, they are hardly used

---

[5] This is often discussed under the keyword 'Mobility as a Service' – MaaS.

for empirical studies and often focus exclusively on the actions of collective actors (state, NGOs, associations) instead of considering the subject's role. The geographer Tim Schwanen and his colleagues (2011) use a governmentality perspective to show that so-called ecological modernization can be understood as a neoliberal project. Referring to Hajer (1997) and some others, they identify dominant logics of the market economy, such as technology optimization, steering the market through prices, and disregarding rebound effects or path dependencies. Governmentality explains the difficulties of integrating alternative forms of knowledge production - beyond the logic of economics, engineering, or psychology - on concrete governmental decisions at the national or local level or on mobility providers (Manderscheid et al. 2014).

3.1 Rationalities, Technologies, and Subjects

Mobility can be understood as a political component of modern societies that (re)produces inequalities and power relations. In our approach, we use the concepts of rationality and technology with references to Foucault and Judith Butler's idea of subject formation.

The term "governmentality" is composed of the concepts of governing ("gouverner") and the way of thinking ("mentalité") (Lemke et al. 2015, p. 8). Governmentality is based on a comprehensive understanding of government and includes additional actors besides the state. Societies seem to govern themselves out of themselves. This does not necessarily happen through direct control or explicit prohibitions but through the ability to induce subjects to act in a certain way and to influence the field of possibilities of individuals (Foucault 1987, p. 255). Discursive necessities are formed, which are internalized by individuals and accepted and desired as guidelines for their own actions. Consequently, power can be found in certain forms of knowledge and truth as well as in the use of technologies of the self (Lemke et al. 2015).

Referring back to Foucault's concept of governmentality (2004a, 2004b), rationalities are described as hegemonic logics of society. These are explicit or implicit logics that influence the individual's way of thinking (Reuber 2012; Lemke et al. 2015). For this article, these can be the rationality of sustainable development, a neoliberal mode of government, or the premise of technological innovation.

Following Foucault, Schwanen et al. (2011, p. 998) describe *techne* as "means, mechanisms, procedures, tactics, vocabularies, etc. [that] are employed to modify the actions of the agents to be governed." The epistemic system describes "which forms of knowledge and expertise are implicated in, constitutive of, and produced by government" (Schwanen et al. 2011, p. 998). Dean (2010, p. 33) also describes this dimension as "specific ways of acting, intervening and directing, made up of particular types of practical rationality ('expertise' and 'know-how') and relying upon definite mechanisms, techniques, and technologies. These concepts are close to the idea of technologies because they conceptualize how the exercise of power works precisely and how theoretical considerations are applied to concrete modes of transport (e.g., shared mobility services). Drawing on Foucault, the concept of technologies describes how governmental goals and logics are translated into regular patterns of action, perception, and judgment. They include material and symbolic instruments, which can act as external technologies or internalize as technologies of the self (Reuber 2012). External technologies in the context of mobility could be obligatory speed limits, traffic lights, or access restrictions for certain vehicles.

The characteristics and effects of technologies and rationalities reveal themselves in the concrete subject formation or subjectification. This describes another element of the governmental exercise of power. The concept of subjectification will be discussed in more detail below.

3.2   Formation of Subjects

In addition to rationalities and technologies, the formation of subjects is also important in sociocultural mobility research. Schwanen et al. (2011, p. 998) address the dimension of *subjectifycation* with the questions "how are the agents to be governed understood, represented and imagined? What are they to become?". Similarly, the consideration of subjects is found in the work of sociologist Katharina Manderscheid (2014). She describes an 'automotive dispositive' that produces individuals as automotive subjects. These are closely interwoven with discourses and collective symbols such as freedom, progress, and individuality. In her work, the mobility dispositive appears as an interplay of complex technologies and material landscapes, forms of knowledge and symbolism, as well as governmental subject formation. Manderscheid (2014, pp. 19f.) emphasizes that a dispositive analytical view of automotive subject formation also includes emotions, preconscious sensations, dispositions, and bodily experiences. Nevertheless, the form in which this can be addressed in empirical research is not explained and is less central concerning her epistemological interest in describing the 'automotive dispositive'.

Subject theorists such as Judith Butler study the formation of subjects more closely. The subject is thus the addressee of regulation. At the same time, subjectification means the constitution of the self via recognition of one's own identity. In this perspective, the incorporation of knowledge and norms leads to a position in which the subject itself influences its own options for action. Foucault's understanding of the subject can be recognized in this double structure:

"There are two meanings of the word 'subject': subject to someone else by control and dependence, and tied to his own identity by a conscience or self-knowledge" (Foucault 1982, p. 781).

However, in contrast to Foucault's genealogical perspective and the sense of the *performative turn*, Butler (2001, 2006) focuses more on the process of identity formation and, with the concepts of performativity and intelligibility, looks at the desire of the subjects.

Butler shares the understanding of power with Foucault and also considers the process of identity formation as an exercise of power (Butler 2006; Reckwitz 2010). Two heuristics are central to the process of identity formation: *Intelligibility* and *Performativity*. She describes the process of the subject becoming intelligible with the idea of invocation, according to Althusser (1977). According to this, a subject becomes intelligible when it can establish a relation between a particular significant (meaning of a linguistic sign) and itself. This act is only about to work if the attribution cited in the significant is accepted and appropriate (Butler 2001). In the process of discursive identity formation, the subject integrates a particular discourse fragment into the view of the self by simultaneously rejecting other identity options in the act of choosing one particular identity component. In Butler's sense, the features to enable a subject's intelligibility are appropriate categories that will allow the subject's positive identification and desire for identification, which Butler describes as desire.

Next to the subject, which becomes intelligible in the act of self-interpretation, Butler adds performativity to analyze identities[6]. In doing so, she refers to the concept of performativity by Derrida and Gasché (1972). By performativity, Butler means the ritualized or habitual citation of speech acts. She argues that the ritualized moment constitutes a "condensed form of historicity" that is "an effect of antecedent and future invocations of convention" (own translation, Butler 2006, p. 12). Identity formation is thereby accomplished through repetition (rite) in everyday use. It is further characterized by convention, the relative independence of time, and the potential for shifting in its content[7].

In the view of urban mobility, the government of subjects as 'traveling bodies' plays an important role (Bonham 2006). Changes or innovations in transport technologies are thus associated (positively or negatively), on the one hand, with the freedom for individuals to move away from social or societal contexts. On the other hand, the understanding of transport as movement from one point to another, to be able to participate in related activities there, enables the objectification of mobility practices. This is accompanied by a corresponding production of knowledge about the efficiency of the movement undertaken. This gives rise to the idea of an 'efficient traveler' or 'efficient body' (Bonham 2006).

3.3 Heuristics for Considering Smart and Shared Mobility Services

After cursorily exploring the concept of governmentality via the concepts of technology, rationality, and subjectification following the approaches of Butler (2006) and Schwanen et al. (2011), the following points crystallize for the empirical illustration:

Rationality

- What logics, strategies, expertise, competencies, and resources are addressed in governmental action?
- How is mobility seen? How is it discursively constituted and justified?

---

[6] The empirical material in this paper does not allow the study of performativity. Instead, we address how subjects are understood as well as represented and with what properties they are constructed.

[7] The two sections on performativity and intelligibility are oriented in an abbreviated form to the subject-theoretical extension of the business power approach, according to Fuchs (2007) in Graf (2016).

Technology

- What means (technologies) are used to translate rationalities into everyday practices and patterns of perception and judgment?
- With what intentions do technologies unfold?
- How is mobility embodied; in terms of bodily attributes and experiences?

Subjectification

- How are subjects understood or represented? With what characteristics are they constructed?
- Which identity options can be identified (intelligibility)?

4 Methodology

This paper combines two methodologies to address the research question and illustrates it with empirical research. We use qualitative content analysis to examine local mobility plans (Schreier 2012; Rädiker and Kuckartz 2019). Local mobility plans are strategic documents of municipalities or regions. They summarize political goals, measures, and indicators in the mobility sector for 10 to 15 years. They provide information about mobility planning as well as the intentions, plans, and strategies of key stakeholders. Consequently, they are a useful source for the empirical illustration of mobility policies. Furthermore, based on a 'mobilized ethnography' (Sheller and Urry 2006; Hein et al. 2008), observation and reconstruction of registration requirements and the conditions for using sharing services are carried out.

The literature corpus consists of four documents in three cities. We got to this corpus by first researching the mobility plans of all sixteen German state capitals. We focused on urban contexts because they appear to be particularly relevant. In cities, companies have begun to launch shared mobility services since densely populated areas represent particularly profitable conditions for sharing services. The development of local mobility plans takes about two (or even more) years. Medard de Chardon (2019, p. 406) describes a "deluge" of free-floating docking-less bike-sharing systems in Europe and North America in 2017, which brought the regulation of shared mobility up on the political agenda, additionally electronic scooters only entered German cities with the Electric Micro-Vehicles Ordinance (eKFV) in 2019. Therefore, we only included documents from before 2019 were not included. In a third step, a lexical search for the terms "sharing", "leih*" (borrow), "miet*" (rent), "geteilt*" (shared) was conducted to identify relevant documents and passages.

Table 1. Selected documents from mobility plan research.[8]

City	Document	Abbreviation	Content
Berlin	Stadtentwicklungsplan Mobilität und Verkehr Berlin 2030 // City Development Plan – Mobility and Transport Berlin 2030 (2021)	B	General mobility plan
Magdeburg	Verkehrsentwicklungsplan 2030plus // Transport Development Plan 2030plus (2019)	MD	General mobility plan
Munich	Mobilitätsstrategie 2035 - Einstieg in die Teilstrategie Shared Mobility // Mobility Strategy – Introduction to the Sub-Strategy on Shared Mobility (2022b)	M	Specific scope on shared mobility (sub-strategy of the general plan)
	Mobilitätsstrategie 2035 - Entwurf einer neuen Gesamtstrategie für Mobilität und Verkehr in München // Mobility Strategy 2035- Draft of a Strategy on Mobility and Transport (2021)	M*	General mobility plan

The analysis of the material is carried out with the analysis software MAXQDA. According to an inductive coding process, the categories are developed directly from the material. Here sequences of the material are analyzed in more detail and assigned to different categories (Rädiker and Kuckartz 2019). The empirical investigation of these plans does not represent a comprehensive analysis of the mobility policy of these cities in terms of a case study but rather serves to approach the research question and illustrate the heuristics developed above.

First, passages with descriptions of subjects and their characteristics, goals, or needs mentioned in connection with mobility were coded. Text passages that describe concrete measures, such as promoting any sharing services, were coded as well. These codes, secondly, allowed inferences to be made about applied technologies. Thirdly, text passages were coded that describe the logic of action of (state) actors, for example, which roles, tasks, and responsibilities are defined and which forms of knowledge are articulated. These codes were then systematized with regard to the research question and based on the developed heuristic of rationalities, technologies, and the process of subjectification.

---

[8] The abbreviations will be used in the following sections to facilitate the reading. The number indicates the page where the quote originates, for example, MD 12 for the document from the city of Magdeburg and page number 12.

The second data collection used for this paper is ethnographic observation (Sheller and Urry 2006; O'Reilly 2012). The registration process of different sharing services was performed exemplary to identify preconditions for using such services. In addition, information was obtained from the general terms and conditions as well as the companies' websites. A total of nine providers were examined. Among them were two car-sharing, three (e-)bike-, one cargo bike-, one electronic moped, and two scooter-sharing providers.[9] The results of this second survey are used in particular for the analysis of the technologies. This enables a more precise understanding of these services beyond the mere mention in mobility plans.

5   Shared Mobility as a Discursive Practice

The following analysis is based on the previously described heuristics for considering mobility transformations from a governmentality perspective. The results of the investigation will be presented based on the concepts of subjectification, technology, and rationality.

All examined mobility plans address shared and smart mobility. The plans of Berlin and Magdeburg are general transport development plans and partly contain targets, indicators, and measures related to general mobility transformation. Shared mobility is dealt with as part of this planning. The city of Munich has also formulated a general strategy with its Mobility Strategy 2035, which includes numerous sub-strategies. The first sub-strategy to be adopted is the Shared Mobility Strategy, which is particularly interesting to the present study.

Consequently, all plans contain general statements promoting smart and shared mobility services. For Berlin, for example, the formulated goal is:

"Strengthening inter- and multimodality and the shared use of vehicles with the aim of a significantly reduced share of MIV in transport" (B 17)[10].

The Munich strategy sets the goal:

"to expand or promote the existing offers city-wide in such a way that they are easily accessible for all and represent a part of everyday mobility for the population" (M 44).

Magdeburg writes under the term Smart Mobility:

"In the future, urban transport should be low-emission and energy-efficient, but also safe, cost-effective, and health-friendly. It is, therefore, not just a matter of increasing digitalization. Rather, a change in mentality and understanding of shared or communally usable and climate-friendly models of locomotion is also crucial" (MD 10).

---

[9] The documentation of this survey can be found in the Annex (see Annex 1). The providers chosen operate in at least one of the cities analyzed.

[10] Originally, this quotation and all following in this section are in German and have been translated by the authors. MIV (German: Motorisierter Individualverkehr) means individual motorized traffic; it includes cars, vans, motorbikes, etc.

5.1 Subjectification

Subjectification describes how subjects are addressed, understood, or represented and what characteristics are attributed to them (cf. section three). The quotes presented above make it clear that all the plans studied take up the concept of shared mobility. However, no specific subjects are addressed. There is only mention of "increased use", accessibility "for all" or "for the population" (see above). Nevertheless, some patterns concerning subjectification become apparent. These will be illustrated under the thematic references of barrier-free accessibility, general subject groups, subjects as citizens in need of protection, and subjects as flexible individuals and consumers.

Barrier-Free Accessibility

Subjects or subject groups with specific characteristics only become apparent in a few places. A central motif is accessibility or the needs of mobility-impaired persons. All cities take up this topic and recognize these groups as subjects. The city of Magdeburg, for example, describes the barrier-free development of the interface between public transport and individual transport (MD 42). The Berlin plan provides for the "establishment of barrier-free accessibility" (B 18) and "equal mobility opportunities for people with mobility impairments" (B 20). At the same time, the plan emphasizes the need for special assistance, which defines a deviation from the 'normal' body and its abilities. Thematically, consideration is given to safety, social participation, and the use of public space or public transport (see, for example, B 20, 27; MD 42, 60, 61; M 48). Consequently, there are many references to accessibility or barrier-free construction, but none discuss the needs of people with reduced mobility with regard to shared mobility access and usability.

General Subject Groups

Other subject groups are addressed, but often in very general collections of identifying characteristics. People of different ages (seniors, children, and teenagers), people regardless of their gender, and social or financial background are listed more than described in their individual needs. According to the Munich Strategy.

> "all individual mobility needs are met quickly, cheaply, and conveniently with a sensible and attractive overall offer. Social backgrounds, age, gender, and physical condition should play no role in this" (M 15).

The other cities have used similar formulations regarding general mobility opportunities (B 20; MD 42, 60).

An exception to these lists is the Munich strategy: here, "spatially but also target group-specific large service gaps" are mentioned, which leads to the fact that "individual service models or products address particular target groups (e.g., tech-savvy young men, or rather above-average earners with a higher level of education)" (M 21). This at least recognizes and describes the unequal use of different subject groups. The strategy does not describe concrete measures or explain how this could be remedied.

Subjects as Citizens in Need of Protection

All cities refer to the 'vision zero' in their general objectives, which states that no more people should be killed or seriously injured in traffic (B 11, 19, 20; MD 43; M* 17f.). In other places, the general safety of all road users is also addressed, which seems to have all users as a whole in mind (B 18, 26; MD 6, 43; M 8, 16). For the city of Magdeburg, the goal is defined as increased "objective and subjective road safety" (MD 43) for all road users. It is thus acknowledged that perceptions of safety underlie subjective interpretations and are not experienced in the same way by all people.

In addition to the understanding of subjects who are in need of safety, health is intensively discussed and increasingly associated with walking and cycling, as with shared mobility (B 11; MD 79). In all plans, healthy conditions of living as well as awareness for issues of health and environmental-friendly behavior are defined as objectives. Health aspects are understood in two ways. On the one hand, environmental impacts, such as traffic emissions (noise, air pollution, etc.) can affect subjects, and on the other hand, subjects themselves can make health-conscious decisions (B 26, 33, 51; MD 10, 26; M* 6, 37)[11]. This can again be seen as a protective measure but also as a hint to people to realize health-conscious lifestyles with the help of shared or active forms of mobility. Health-conscious actions are linked in the plans to the identity proposition of a healthy lifestyle. In both themes, it is implicitly the motorized individual transport that must be overcome to realize (public) health.

In addition, there is the role of the subjects as democratic citizens. In the Berlin Plan, for example, public space is described "as a focal point of public life" (B 27). Mobility is understood as an opportunity to participate in public life. Further, the quality of stay in urban space is emphasized (B 20, 33; M 11, 16). The Munich Strategy, for example, describes the conversion of vacant areas of stationary traffic to increase the quality of stay in public spaces (M 11). In addition, the Munich strategy emphasizes the acceptance of the citizens. In this respect, additional reference is made to the (democratic) legitimacy of decisions, on which all planning and political decisions should measure their quality (B 10, 16; M 11, 35, 57). The linking of mobility transformations with questions of political participation can thus be found rudimentarily in the plans but is not yet sophisticated.

Subjects as Flexible Individuals and Consumers

Individuality and flexibility are prominently described as necessary resources for the subjects. "Individual direct connections" as well as "flexible intermediate stops" (M 10) are decisive criteria for the use of different mobility options. A broad vehicle portfolio in shared mobility would enable more individualization and flexibility (M 15). Comfort, reliability, and privacy are described as further needs of mobile subjects (B 15; M 12, 25, 26, 38, 47). Subjects are thus described in terms of an 'efficient body' (Bonham 2006). For the Shared Mobility Strategy of Munich, a strong link between the role of citizens and users becomes clear. The strategy pursues "as its highest priority a strong orientation

---

[11] The second reading could also be understood in terms of self-technology. Insofar as people take up the rationality of a health-conscious lifestyle and translate it with the help of shared or active forms of mobility, governmental governance would emerge here. However, an association with health-conscious actions is increasingly associated in the plans with walking and cycling, rather than shared mobility (B 11; MD 79).

towards people as citizens and users" (M 15). In addition, subjects are described as the target group of new users of sharing services (M 15, 21, 28, 49, 51). From this, an understanding of citizens as so-called *citizen-consumers* can be read. In addition to their role as citizens in need of protection, citizens are also understood as consumers whose consumption decisions impact urban mobility. In this context, the factors of individual and flexible use of transport modes are mentioned as decision criteria. The possibility of choice and the power of individuals to decide are equally emphasized here and can ultimately be traced back to liberal notions of freedom.

5.2 Technologies

Technologies are instruments to realize everyday actions, perceptions, and judgments. Technologies can pursue different intentions, forms, or strategies. In the following, various forms of technologies will be outlined in the sense of an exemplary clarification from observation.

All the cities studied take up the facilitation of sharing offers in their mobility planning. The role of city administration and politics is predominantly seen in creating appropriate regulations for sharing services. This includes providing space and creating incentives for users and companies (B 32; MD 67, 71, 72; M 7, 15, 16, 17, 51). In Munich, for example, a "'level playing field' for non-discriminatory and fair competition" is to be created (M 16). Nonetheless, there is also the approach of intervening in a regulatory manner if supply gaps open up (see above). The operational business of shared mobility services is then no longer the responsibility of municipal institutions but of private companies. The provision of (public) space for private entrepreneurial use of mobility services also promotes the commodification of public space.

The exemplary observation of shared mobility registration processes (see Table 1 in the appendix) shows their usage requirements. A mail address and some form of proof of identity are required for all services. Using the service without personalized registration is impossible, as one can with buses or trains. Except for a cargo bike rental and a local car-sharing provider, a smartphone with mobile data and GPS function is required. For the smartphone, the usage of an app is foreseen, which can be downloaded via an Apple ID or PlayStore ID. Alternative operating systems are accordingly not supported for these services. Almost all the services examined are operated commercially, so a credit card or online payment service (Paypal, Apple Pay) must be used. The use of online payment services, in turn, requires certain liquidity and usually the existence of a bank account. For the operation of motorized modes, such as the car, a driver's license must also be available. The official age limit for using the mobility options is 18 years. The eKFV allows usage at the age of 14 years. Providers implemented higher age limits due to insurance coverage and reliability.

In addition to these formal access requirements, additional skills that are needed can lead to exclusionary dynamics. For example, in addition to owning an appropriate smartphone, one must also be proficient in using it (see also Groth 2019). The actual use of the services requires physical as well as psychological skills. Micro-mobility services can also be used for individual purposes. For example, car-sharing offers do not include child seats, which means that families or people providing care work can only use these offers with considerable additional organizational effort. Similar hurdles arise for bike-sharing. The bicycles are standardized one-size-fits-all and are therefore only aimed at people within a certain physical norm. Especially with free-floating services, there is often competition for space on sidewalks (B 26; MD 60f., 79), so that especially those who walk are negatively affected. From a statistical point of view, in Germany, this is mainly the case for children up to 9 years of age, as well as people over 70 years of age, and more women than men (Nobis and Kuhnimhof 2018). Initial studies on sharing service users show that mainly young, male, and above-average educated residents of urban areas use these services (Médard de Chardon 2019; Laa and Leth 2020; Pangbourne et al. 2020; Reck et al. 2022). Regarding the concept of intelligibility, these mobility services seem appropriate for specific user groups only. Subjects outside this group seem to regard different sharing offers as less or not at all appropriate offers.

5.3 Rationalities

Rationalities are forms of knowledge and representations that implicitly presuppose or (re)produce governmental action. All cities refer to the concept of sustainability in their mobility plans. Often the connection with climate protection goals, as well as the promotion of the environmental alliance, is mentioned (B 6, 17, 20, 24; MD 44, 57, 121; M 11, 18, 44; M* 3). The Munich strategy establishes a direct link between shared mobility and climate neutrality:

> "Shared Mobility actively contributes to the achievement of city-wide climate neutrality and becomes exclusively climate-neutral and low-emission by 2035" (M 18).

The plans of Berlin and Magdeburg make this connection less explicit. Nevertheless, the promotion of shared mobility is also included as a measure in the plans here (see the section on technologies).

Furthermore, all plans apply the standards of efficiency and profitability to the mobility system. Thus, the guarantee of an "attractive door-to-door travel time" (MD 42), increased efficiency and interconnectedness in the transport sector, profitability as well as the functioning of commercial transport are formulated as demands or goals (B 50; MD 38, 42, 57, 59; M 16, 17, 44, 47, 49). This logic implies solving problems in the mobility sector by creating more alternative and efficient options without problematizing environmentally harmful and unequal forms of mobility comparably.

Another strategy can be seen in the attempt to upgrade public space with sustainable mobility. In this perspective, sustainable mobility virtually pays for the attractiveness of locations because new businesses are established, or more areas are freed up for greening. The Berlin Mobility Plan states: "Berlin is an attractive market for new (mobility) offers" (B 6). The Munich strategy, for example, describes the conversion of vacant areas of stationary traffic for the benefit of the quality of stay in public space or to enable new offers for shared mobility (M 11). This reading brings the efficient use of public space and its commodification back into focus.

6 Conclusion

Current transport policy emphasizes the role of shared mobility services, such as car- and bike-sharing, as well as shared e-scooters and mopeds-. With the help of a governmentality perspective, promoting these services can be understood as governance impulses for society and individuals alike. With the content analysis of different local mobility plans and the observation of registration requirements for sharing services, the theoretical concepts of subjectification, technology and rationality could be applied to practical mobility plans and shared mobility services. The results show that the constitution of individual and flexible citizen-consumers corresponds with the rationality of a social and economic structure oriented towards (green) economic growth. Rationalities such as fair competition or locational advantages through sustainable development further underline this impression.

The promotion of shared mobility services unfolds its governmental power in two ways: On one hand, they enable subjects to behave according to a sustainable and smart mobility transformation. On the other hand, they imply certain preconditions for use. Governmental regulation of shared mobility services focuses on providing and enabling additional services so that subjects are guided to use them. The identity formations for a healthy, environmentally friendly, modern transport behavior are suitable for shaping individuals' desires and are equally suitable for municipalities' efforts towards sustainable and smart transformation. However, the analysis of local mobility plans and the observation of registration requirements of micro-mobility allow only a limited perspective on subjectification processes. The extent to which individuals accept these identification offers (performativity), for example, seeing themselves as citizens in need of protection or as citizen-consumers, cannot be answered within the framework of this evaluation. This could be explored, for example, by conducting (narrative) interviews.

Our explanations show that shared mobility services are not (or cannot be) used equally by everyone. A twofold inequality accompanies this. On the one hand, existing disparities in mobility behavior are not addressed accordingly, and, on the other hand, inequalities are partly reproduced and consolidated. Shared mobility services are primarily aimed at people who conform to physical and social majority norms - for example, in the case of shared bicycles. It is also necessary to have a smartphone that can be operated and used. Finally, the physical prerequisites in the sense of a certain age and physical and mental abilities are needed to enable legal and unproblematic use. In light of the construction of citizen-consumers, this seems particularly relevant: People who, for whatever reason, do not appear as users of shared mobility services may no longer be perceived as stakeholders, so their interests are easily pushed to the background or get overlooked (see also Kronsell and Mukhtar-Landgren 2020). Thus, mobility as a social and democratic issue becomes more urgent. The purely quantitative increase of mobility services can only break up the existing inequality to a limited extent. Without accompanying measures for barrier-free and affordable access to mobility services and regulating negative externalities, smart and shared mobility services threaten to remain trapped in a (neoliberal) logic of growth. To put this provocatively: People with high potential for mobility gain additional mobility options through shared vehicles, whereas people who are already threatened or affected by immobility seem to be (still) denied access to new mobility services.

The impulses for the transformation of the mobility system, such as changed legislation regarding accessibility and shared mobility, digitalization, or intensified climate policies outlined at the beginning, are relatively new. They require expanding mobility services for less mobile subjects to catch up with the average mobility level. On the other hand, developing comprehensive local mobility plans can take several years. It is, therefore, not possible or meaningful to make a conclusive assessment at this stage.

Concerning shared mobility services in our cases, policies and governmental use of power are revealed mainly through the support of technical innovations as well as the creation of fair competition among different providers. Additionally, the enabling role of the state is emphasized but often only manifests itself in the commodification of public space. This enabling role of policy is potentially multifaceted. In various urban contexts, increasing regulation of free-floating shared mobility services can be seen. For example, no-parking zones can be defined, a proof-of-parking picture can be required, and clear parking facilities for micro-mobility can be created in (car) parking areas to reduce thus conflicts on curbsides (Marsden et al. 2020; Munich 2022a). Possible conclusions from this could also be a stronger focus on diversified forms of shared mobility. For example, shared cargo bikes or car-sharing with child seats could enable additional uses. Locally or publicly funded and/or supported sharing operators can offer lower-threshold services. In addition, driving training or the integration of underrepresented user groups can help to make the services available to marginalized subjects. In addition, all forms of shared micro-mobility depend on appropriate transport infrastructure. Thus, walking and cycling paths, in particular, are used by these mobility modes. A consistent expansion of these paths and decelerating road traffic should be additional supporting measures in future mobility development plans.

References

Althusser, L.: Ideologie und ideologische Staatsapparate: Aufsätze zur marxistischen Theorie. Hamburg (1977)

Banister, D.: The sustainable mobility paradigm. Transp. Policy **15**, 73–80 (2008). https://doi.org/10.1016/j.tranpol.2007.10.005

Berlin. Stadtentwicklungsplan Mobilität und Verkehr Berlin 2030 (2021). https://www.berlin.de/sen/uvk/verkehr/verkehrspolitik/stadtentwicklungsplan-mobilitaet-und-verkehr/. Accessed 29 June 2022

Bonham, J.: Transport: disciplining the body that travels. In: Paterson, M., Boehm, S. (eds.) Against Automobility, pp. 57–74. Hoboken, NJ, Oxford (2006)

Butler, J.: Das Unbehagen der Geschlechter, 5th edn. Frankfurt am Main (1995)

Butler, J.: Psyche der Macht: Das Subjekt der Unterwerfung. Frankfurt am Main (2001)

Butler, J.: Haß spricht: Zur Politik des Performativen. Frankfurt am Main (2006)

Cresswell, T.: Towards a politics of mobility. Environ. Plann. D Soc. Space **28**, 17–31 (2010). https://doi.org/10.1068/d11407

Dean, M.: Governmentality: Power and Rule in Modern Society, 2nd edn. London (2010)

Deffner, J.: Zu Fuß und mit dem Rad in der Stadt: Mobilitätstypen am Beispiel Berlins (Dortmunder Beiträge zur Raumplanung 7) (2009)

Derrida, J., Gasché, R.: Die Schrift und die Differenz. Frankfurt am Main (1972)

Docherty, I., Marsden, G., Anable, J.: The governance of smart mobility. Transp. Res. Part A Policy Pract. **115**, 114–125 (2018). https://doi.org/10.1016/j.tra.2017.09.012

European Commission. Handbook on the external costs of transport. Luxembourg (2019)

European Commission. Sustainable and Smart Mobility Strategy –putting European transport on track for the future: COM (2020). 789 final

European Commission. The New EU Urban Mobility Framework: Communication from the Commission to the European Parliament, the Council, the European Economic and Social Committee and the Committee of the Regions. COM (2021). 811 final

European Environment Agency. The first and last mile: The key to sustainable urban transport: transport and environment report 2019 (EEA report 18). Luxembourg (2019)

Foucault, M.: The subject and power. Crit. Inq. **8**(4), 777–795 (1982)

Foucault, M.: Das Subjekt und die Macht. In: Dreyfus, H.L., Rabinow, P. (eds.) Michel Foucault. Jenseits von Strukturalismus und Hermeneutik, pp. 243–264. Frankfurt am Main (1987)

Foucault, M.: Die Geburt der Biopolitik: Geschichte der Gouvernementalität II. Vorlesung am Collège de France 1978–1979. Frankfurt am Main (2004a)

Foucault, M.: Sicherheit, Territorium, Bevölkerung: Geschichte der Gouvernementalität I. Vorlesung am Collège de France 1977–1978. Frankfurt am Main (2004b)

German Federal Government. Klimaschutzprogramm 2030 (2019). https://www.bundesregier ung.de/breg-de/themen/klimaschutz/klimaschutzprogramm-2030-1673578. Accessed 23 June 2022

Gössling, S., Cohen, S.: Why sustainable transport policies will fail: EU climate policy in the light of transport taboos. J. Transp. Geogr. **39**, 197–207 (2014). https://doi.org/10.1016/j.jtrangeo. 2014.07.010

Graf, A., Sonnberger, M.: Responsibility, rationality, and acceptance: how future users of autonomous driving are constructed in stakeholders' sociotechnical imaginaries. Public Underst. Sci. **29**, 61–75 (2020). https://doi.org/10.1177/0963662519885550

Groth, S.: Multimodal divide: Reproduction of transport poverty in smart mobility trends. Transp. Res. Part A: Policy Pract. **125**, 56–71 (2019). https://doi.org/10.1016/j.tra.2019.04.018

Hajer, M. A.: The Politics of Environmental Discourse: Ecological Modernization and the Policy Process. Oxford (1997)

Hein, J.R., Evans, J., Jones, P.: Mobile methodologies: theory, technology and practice. Geogr. Compass **2**, 1266–1285 (2008). https://doi.org/10.1111/j.1749-8198.2008.00139.x

Heinen, E., Mattioli, G.: Multimodality and CO2 emissions: a relationship moderated by distance. Transp. Res. Part D: Transp. Environ. **75**, 179–196 (2019). https://doi.org/10.1016/j.trd.2019. 08.022

Horjus, J.S., Gkiotsalitis, K., Nijënstein, S., Geurs, K.T.: Integration of shared transport at a public transport stop: mode choice intentions of different user segments at a mobility hub. J. Urban Mob. **2**, 100026 (2022). https://doi.org/10.1016/j.urbmob.2022.100026

Kronsell, A., Mukhtar-Landgren, D.: Experimental governance of smart mobility: some normative implications. In: Paulsson, A., Hedegaard Sørensen, C. (eds.) Shaping Smart Mobility Futures: Governance and Policy Instruments in Times of Sustainability Transitions, pp. 119–135. Bingley (2020)

Laa, B., Leth, U.: Survey of E-scooter users in Vienna: who they are and how they ride. J. Transp. Geogr. **89**, 1–8 (2020). https://doi.org/10.1016/j.jtrangeo.2020.102874

Lemke, T., Krasmann, S., Bröckling, U.: Gouvernementalität, Neoliberalismus und Selbsttech-nologien. Eine Einleitung. In: Bröckling, U., Krasmann, S., Lemke, T. (eds.) Gouvernemental-ität der Gegenwart: Studien zur Ökonomisierung des Sozialen, pp. 7–40. Frankfurt am Main (2015)

Lubitow, A., Abelson, M.J., Carpenter, E.: Transforming mobility justice: gendered harassment and violence on transit. J. Transp. Geogr. **82**, 1–7 (2020). https://doi.org/10.1016/j.jtrangeo. 2019.102601

Lucas, K.: Transport and social exclusion: where are we now? Transp. Policy **20**, 105–113 (2012). https://doi.org/10.1016/j.tranpol.2012.01.013

Lyons, G.: Getting smart about urban mobility – aligning the paradigms of smart and sustainable. Transp. Res. Part A: Policy Pract. **115**, 4–14 (2018). https://doi.org/10.1016/j.tra.2016.12.001

Magdeburg. Verkehrsentwicklungsplan 2030plus. Magdeburg (2019). https://www.magdeb urg.de/Start/B%C3%BCrger-Stadt/Leben-in-Magdeburg/Planen-Bauen-Wohnen/Verkehrse ntwicklungsplan-2030plus.php?ModID=10&FID=37.821.1&object=tx%7C37.14051.1&red ir=1. Accessed 29 June 2022

Manderscheid, K.: Formierung und Wandel hegemonialer Mobilitätsdispositive. Zeitschrift Diskursforschung (1), 5–31 (2014)

Manderscheid, K., Schwanen, T., Tyfield, D.: Introduction to special issue on 'mobilities and Foucault'. Mobilities, **9**, 479–492 (2014). https://doi.org/10.1080/17450101.2014.961256

Marsden, G., Docherty, I., Dowling, R.: Parking futures: curbside management in the era of 'new mobility' services in British and Australian cities. Land Use Policy **91**, 1–10 (2020). https://doi.org/10.1016/j.landusepol.2019.05.031

Marsden, G., Reardon, L.: Introduction. In: Marsden, G., Reardon, L. (eds.) Governance of the Smart Mobility Transition, pp. 1–15. Bingley (2018)

Mattioli, G.: Transport needs in a climate-constrained world. A novel framework to reconcile social and environmental sustainability in transport. Energy Res. Soc. Sci. **18**, 118–128 (2016). https://doi.org/10.1016/j.erss.2016.03.025

Mattioli, G., Heinen, E.: Multimodality and sustainable transport: a critical perspective. In: Appel, A., Scheiner, J., Wilde, M. (eds.) Mobilität, Erreichbarkeit, Raum. SMV, pp. 65–82. Springer, Wiesbaden (2020). https://doi.org/10.1007/978-3-658-31413-2_5

May, A.D.: Urban Transport and sustainability: the key challenges. Int. J. Sustain. Transp. **7**, 170–185 (2013). https://doi.org/10.1080/15568318.2013.710136

Médard de Chardon, C.: The contradictions of bike-share benefits, purposes and outcomes. Transp. Res. Part A: Policy Pract. **121**, 401–419 (2019). https://doi.org/10.1016/j.tra.2019.01.031

Mullen, C., Marsden, G.: Mobility justice in low carbon energy transitions. Energy Res. Soc. Sci. **18**, 109–117 (2016). https://doi.org/10.1016/j.erss.2016.03.026

Munich. Mobilitätsstrategie 2035 - Entwurf einer neuen Gesamtstrategie für Mobilität und Verkehr in München: Beschluss über die Finanzierung ab 2021. Sitzungsvorlagen Nr. 20–26/V 03507 (2021)

Munich. Evaluierung der verkehrlichen Wirkungen von E-Tretrollern (2022a). https://muenchenu nterwegs.de/content/1423/download/220530-bericht-eva-et-final-web.pdf. Accessed 29 June 2022

Munich. Mobilitätsstrategie 2035 Einstieg in die Teilstrategie Shared Mobility: Etablierung von Mobilpunkten und Angebotsausweitung in München. Sitzungsvorlage 20–26/V 04857 (2022b)

Nobis, C., Kuhnimhof, T.: Mobilität in Deutschland: Ergebnisbericht (2018). http://www.mobili taet-in-deutschland.de/pdf/MiD2017_Ergebnisbericht.pdf. Accessed 28 June 2022

O'Reilly, K.: Ethnographic Methods, 2nd edn. Abingdon, Oxon, New York (2012)

Pangbourne, K., Mladenović, M.N., Stead, D., Milakis, D.: Questioning mobility as a service: unanticipated implications for society and governance. Transp. Res. Part A: Policy Pract. **131**, 35–49 (2020). https://doi.org/10.1016/j.tra.2019.09.033

Paulsson, A., Hedegaard Sørensen, C. (eds.): Shaping Smart Mobility Futures: Governance and Policy Instruments in Times of Sustainability Transitions. Bingley (2020)

Priya Uteng, T.: Gender, ethnicity, and constrained mobility: insights into the resultant social exclusion. Environ. Plann. A: Econ. Space **41**, 1055–1071 (2009). https://doi.org/10.1068/a40254

Rädiker, S., Kuckartz, U.: Analyse qualitativer Daten mit MAXQDA. Wiesbaden (2019)

Reck, D.J., Martin, H., Axhausen, K.W.: Mode choice, substitution patterns and environmental impacts of shared and personal micro-mobility. Transp. Res. Part D: Transp. Environ. **102**, 103134 (2022). https://doi.org/10.1016/j.trd.2021.103134

Reckwitz, A.: Subjekt, 2nd edn. Bielefeld (2010)

Reuber, P.: Politische Geographie. Paderborn (2012)

Schreier, M.: Qualitative Content Analysis in Practice, 2012 edn. London (2012)

Schwanen, T., Banister, D., Anable, J.: Scientific research about climate change mitigation in transport: a critical review. Transp. Res. Part A: Policy Pract. **45**, 993–1006 (2011)

SHARE-North. A Planner s Guide to the Shared Mobility Galaxy (2021). https://share-north.eu/the-guide/. Accessed 17 Oct 2022

Sheller, M.: Theorising mobility justice. Tempo Soc. **30**, 17–34 (2018). https://doi.org/10.11606/0103-2070.ts.2018.142763

Sheller, M., Urry, J.: The new mobilities paradigm. Environ. Plann. A: Econ. Space **38**, 207–226 (2006). https://doi.org/10.1068/a37268

Sheller, M., Urry, J.: Mobilizing the new mobilities paradigm. Appl. Mob. **1**, 1–16 (2016). https://doi.org/10.1080/23800127.2016.1151216

Sonnberger, M., Graf, A.: Sociocultural dimensions of mobility transitions to come: introduction to the special issue. Sustain. Sci. Pract. Policy **17**, 174–185 (2021). https://doi.org/10.1080/15487733.2021.1927359

Framing Digital Mobility Gap: A Starting Point in the Design of Inclusive Mobility Eco-Systems

Nina Nesterova[1]([✉]), L. Hoeke[2], J. A.-L. Goodman-Deane[3], S. Delespaul[4], Bartosz Wybraniec[5], and Boris Lazzarini[5]

[1] Research and Business Innovation, Academy of Built Environment and Logistics, Breda University of Applied Sciences, Breda, The Netherlands
Nesterova.N@buas.nl
[2] Breda University of Applied Sciences, Breda, The Netherlands
[3] University of Cambridge, Cambridge, UK
[4] Mobiel 21, Leuven, Belgium
[5] The Polytechnic University of Catalonia, Barcelona, Spain

Abstract. Digital transport eco-systems worldwide provide great advantages to many but also carry a risk of excluding population groups that struggle with accessing or using digital products and services. The DIGNITY project (DIGital traNsport In and for socieTY) delves into the development of such eco-systems to deepen the understanding of the full range of factors that lead to disparities in the uptake of digital transport solutions in Europe. A starting point for developing digitally inclusive transport systems is to obtain state-of-the-art knowledge and understanding of where local transport eco-systems are in relation to the digital gap and digital mobility gap in terms of their policies, transport products and services, and population digital literacy. This chapter presents the methodology developed in the DIGNITY project to frame this digital gap, incorporating a self-assessment framework that may be used by public authorities to identify potential gaps in the development of local digital transport eco-systems. This framework is informed by results from customer journey mapping exercises that provide insights into the daily activities and trips of users, and larger scale surveys on digital technology access, use, attitudes and competence in the area. In the DIGNITY approach as a whole, the results from the framing phase are then used to inform subsequent work on bridging the digital gap through the co-creation of more inclusive policies, products and services. The chapter provides concrete results from the framing exercise in four DIGNITY pilot areas: Barcelona, Tilburg, Flanders and Ancona. The results clearly show that a digital transport gap exists in these areas, and that this is manifested in different ways in different local situations, requiring tailored approaches to address the gap.

1 Introduction

"Transitioning from a paper card to a chipcard to use public transport, gave me a lot of anxiety. It took me a long time to learn where to hold my card against the machine to validate my ride, which embarrassed me towards other passengers" says an elderly woman. A low-income, migrant woman mentions: "I buy tickets at the station, because I

© The Author(s) 2023
I. Keseru and A. Randhahn (Eds.): *Towards User-Centric Transport in Europe 3*, LNMOB, pp. 235–253, 2023.
https://doi.org/10.1007/978-3-031-26155-8_14

don't have an online account to buy them. I don't like to pay by mobile because I'm not very tech savvy". Visually impaired man states: "With apps, you cannot zoom in and I do not always carry glasses with me. Then I use voiceover. I can use it, but I do not like it personally". An elderly couple share their experience: "I would rather do it by phone because I'm afraid I'll make a mistake and give them too much money; I am not confident enough on the computer. I just prefer to speak to someone" (Nesterova et al. 2021). While digitalisation of different economy sectors and transition to smart cities are becoming our everyday reality, there is also a growing concern that the fast digitalisation pace leads to disparities in the uptake of digital transport solutions within different population groups in Europe, becoming a new risk factor for transport poverty. Public authorities are faced with a challenge to combine the opportunities from the digitalisation of the mobility eco-system with problems arising from this process. Banister (2019) says: "As with many innovations that have huge potential to benefit all society, it is the rich and those with the necessary knowledge and supporting infrastructure who are the main gainers. However, if the objectives of transport policy are to reduce levels of relative and absolute inequality, then priority needs to be given to providing the means by which all members of society can benefit from innovation". A starting point for the development of the digitally inclusive transport systems is to obtain state-of-the-art knowledge and understanding of where local transport eco-systems are in relation to the digital gap and digital mobility gap in terms of their policies, transport products and services and population digital literacy.

The DIGNITY project (DIGital traNsport In and for socieTY) delves into the development of digital mobility eco-systems and contributes to the better understanding of the full range of factors that lead to disparities in the uptake of digital transport solutions in Europe. Financed under European Union's Horizon 2020 research and innovation programme, DIGNITY brings together the partners from 6 countries, to analyse the digital transition from user and provider's perspective and to design, test and validate a novel concept for development of the digital inclusive travel system. DIGNITY approach is developed and validated within four pilots:

- The pilot in Ancona, the capital city of the Marche region (Italy) with less 100000 inhabitants.
- The pilot in Barcelona, embracing a population of 1.7 million inhabitants.
- The pilot region of Flanders, with population of around 6.5 million inhabitants.
- The pilot in Tilburg, a city located in the south of the Netherlands, which counts 217595 inhabitants.

This chapter provides an overview of the methods available for the public authorities to frame the digital mobility gap in their region, as a starting point in the development of the inclusive mobility eco-systems.

2   Framing the Digital Gap

Design of the inclusive mobility eco-system requires an integral approach that brings together needs, attitudes and requirements of the transport stakeholders on micro, meso and macro levels:

- The Micro level of the DIGNITY comprises all citizens and all possible users of digital mobility products and services.
- The Meso level of the DIGNITY is about the digital mobility products and services available within a region.
- At the Macro level, the institutional structure of a region is considered (political administration, as well as other forms of political regulation, network governance and the interdependence of political decision-making levels).

Integrating these three levels into one holistic DIGNITY methodology, it proposes to take a three-phase approach (Fig. 1). Within this approach the "framing phase" creates an understanding of how much the digital divide contributes to the mobility poverty of different population groups; "bridging phase" focuses on co-creation of the solutions for design of more inclusive transport policies, products and services; and "evaluation phase" looks at impacts of the overall process and ensures contribution to the formulation of the long-term strategies to fill in the gap.

Fig. 1. DIGNITY approach to the development of the inclusive mobility eco-system.

This chapter presents the "framing phase" of the methodology. It suggests to start with the self-assessment framework that allows public authorities to identify potential gaps in the development of local inclusive digital transport eco-systems. Framing phase looks at how many people are at risk of being excluded in the mobility sector and why. By analysing how and why target groups are using (or not) existing products and services, a more detailed understanding of vulnerable users and their needs is created. Thus, the objective of the framework is to support public and private mobility providers in generating a knowledge; where current digital transport systems risk leaving some population groups behind, and in conceiving mainstream digital products or services that are accessible to and usable by as many people as possible, regardless of their income, location, social or health situation or age. The framework is informed by results from customer journey mapping exercises that provide insights into the daily activities and trips of users, larger scale surveys on digital technology access, use, attitudes and competence in the area, and focus groups zooming into the needs and attitudes of the specific population groups, further detailed in the following paragraphs. Results from the framing phase provide an overall understanding of the digital gap in mobility, allowing to zoom into the different stakeholder groups and getting a more in-depth knowledge about challenges of each. Within DIGNITY approach as a whole, the results from the framing phase are then used to inform subsequent work on bridging the digital gap through the co-creation of more inclusive policies, products and services.

3   Building Blocks of the Framing Phase

DIGNITY framing phase includes four distinctive methods:

- The digital gap self-assessment
- Customer Journey Mapping
- Large Scale Surveys
- Focus groups.

3.1   The Digital Gap Self-assessment in Mobility

The digital gap self-assessment framework provides cities and regions with a clear representation of the digital gap in mobility in their region. It includes:

- the knowledge about digital abilities and mobility of citizens;
- an overview of the current market supply of digital mobility products and services;
- and the policy readiness to act on digitalisation in mobility.

Performing the digital gap self-assessment helps public authorities to identify focus areas in their policy making processes.

The method combines in one comprehensive framework the state of mobility digitalisation at three DIGNITY levels (Fig. 2). Each of these levels is further detailed in the groups of indicators (composed of the detailed indicator set):

- Micro level indicator groups are: digitalisation in mobility; population; mobility; digital ability.
- Meso level indicator groups are: usage by vulnerable groups; stakeholders; digital transport provision.
- Macro level indicator groups are: government structures; regulatory framework; budget and outreach programs.

The self-assessment method offers a description of these indicators, possibilities for data collection methods and provides an Excel-sheet to fill in the collected information. The methods for data collection are, for example, the use of national or local statistics, population surveys (micro-level), cross-department working or focus groups within public authorities, questionnaires towards mobility providers (meso-level); interviews within public authorities and focus groups (macro-level).

Establishment of the links between micro, macro and meso levels (Fig. 2) allows to identify the areas where potential gaps in the inclusive mobility eco-system exists. For example, combining micro and meso level data makes visible the supply (or lack

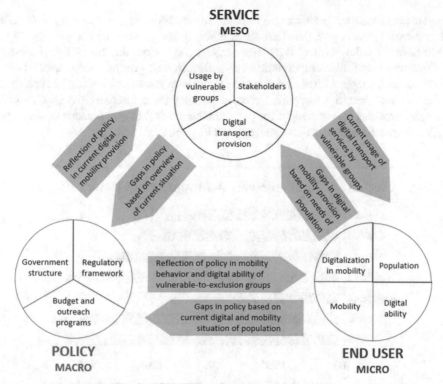

Fig. 2. DIGNITY self-assessment framework

of it) of digital mobility products/services to specific vulnerable to exclusion group. It becomes clear if a group is underrepresented in the use of a product/service and where inclusive design has a potential to improve it, making it accessible to larger population groups. Bringing the information of meso and macro level together, can help to identify the gaps in policy and regulation necessary to create an inclusive mobility eco-system. Finally, the micro level data provides important information for local authorities at the macro level. Combining data from these two levels allows to identify which vulnerable to exclusion group experiences mobility poverty the most, how big is this problem and where a dedicated regulative and institutional support is the most urgent.

All DIGNITY pilots have performed the self-assessment for their municipality/region as the first step in the framing phase. Within the evaluation phase they expressed that this method was very structured, detailed and sometimes too complex. Improvement can be achieved in better guiding the stakeholders in the data and information to collect and in the advice on how to combine different information sources. A more flexible and less structured method would be more beneficial in some cases. Overall, the self-assessment data collection process allowed to realise what data types are missing in local context to create a full understanding on the scope and size of the digital mobility gap problem.

Results for this self-assessment methods provide public authorities an overall understanding of the size of the digital gap in mobility, allowing to zoom into the different assessment levels and getting more in-depth information for each of the levels. With this, potential gaps in policy and supply of mobility services can be derived. The Fig. 3 below shows the data, collected by large scale surveys, from the use of digital services in Flanders among different vulnerable-to-exclusion user groups (micro level). In this graph, elderly, people with disabilities and people with low education can be considered as the less confident to plan a public transport journey using internet or an app. This, for example, provides a clear insight for policy makers that this group needs another than digital approach (or support in digital approach) within this activity.

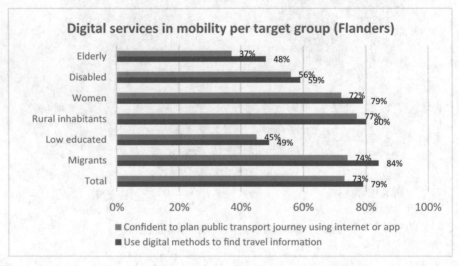

Fig. 3. Example of the self-assessment result: digital services in mobility per target group in Flanders.

Another example is shown in Table 1 where the list of main mobility services and products in Barcelona are identified (meso level). This shows the diversity of the market supply of digital mobility products and illustrates that alternative to the digital version is not always available on the market, meaning that some vulnerable to exclusion user groups are partially excluded from this mobility option. Combination of interviews; literature review and media review was used to collect this information.

On the macro-level, self-assessment framework results have indicated the pilots the readiness or unreadiness of the institutional and regulatory system for the digitalisation in mobility. For example, it was even difficult to find specific references to digital mobility in policy documents, showing unexplored potential for departments to be involved in the digitalisation of transport. In Tilburg, cross disciplinary collaboration is already taking place and vulnerable-to-exclusion groups are involved in policy developments. However, it is acknowledged that the complexity of the government structure results in a barrier for certain groups in Tilburg to be well involved and represented in the decision-making process on inclusive mobility.

Table 1. List of main mobility services and products in Barcelona.

Category of digital service or product	Number of services and examples	Non-digital alternatives for product or service
Trip planning	Around 10: City trips, google maps…	Static information of bus lines
Navigation	5 to 10: HERE, INRIX, Garmin…	Paper maps
Parking payment	5 to 10: Wesmartpark, Via-T, Smou,…	Non-digital parking payment
Consumer car sharing	Less than 5: Ubeeqo, Virtuo	No non-digital alternative
Personal car sharing	Less than 5: Social Car, Getaround	No non-digital alternative
Corporate car sharing	Less than 5: Ubeeqo	No non-digital alternative
Ride splitting	Less than 5: Blablacar, Amovens, Journify, RACC Hop	No non-digital alternative
E-hailing (taxis)	Less than 5: Cabify, Social Car	Taxi
Demand responsive public transport	Less than 5: Shotl, Ne-MI	No non-digital alternative
Bike sharing	Less than 5: Bicing, Donkey Republic, Mobike	No non-digital alternative
Other vehicle sharing	Scooters around 5: Yego, Cooltra, SEAT MÓ, Acciona, Movo, Gecco Kick scooters less than 5: Reby	No non-digital alternative
Vehicle information	10 to 15: Google maps, apps of mobility services…	No non-digital alternative
Parking information	Around 5: Parkopedia, Parclick, Telpark, Wesmartpark	No non-digital alternative
Facility information	Charging station apps 5 to 10: AMB-electrolineres, Charge Map, Plug share. Bike stations less than 5: Google maps, City trips	No non-digital alternative
Travel information	5 to 10: Waze, Mou-te, RACC…	No non-digital alternative
Roadside assistance	Less then 5: RACC assistència	No non-digital alternative

3.2 Customer Journey Mapping

Customer Journey Mapping (CJM) is a method known in marketing to map and measure the experiences of users in the form of micro-scale qualitative data. In DIGNITY, it is used to understand mobility challenges of selected vulnerable to exclusion groups.

The method allows to pinpoint specific problems and issues in a predefined journey that the user will make. Collecting these insights for a specific situation or journey results in a clear improvement potential for the mobility products/services that are needed to increase the overall journey experience of the end user.

The CJM method consists in several steps:

- Define the journey;
- Define the target group;
- Define activities and touchpoints of the journey;
- Prepare research: recruit participants and prepare questions;
- Execute the research and analyses the data.

The chosen journey can be derived from the outcomes of the self-assessment method (e.g. potential gaps identified from meso and micro data) or be suggested by local industry or policy-making stakeholders. This can, for example, be a bus trip from home to the train station; the use of a new mobility service (e.g. a shared car); the use of a navigation app or buying a ticket at a ticket machine. The CJM method focuses on the moments that the participant interacts with the mobility product/service, measuring the experience of the participant during these moments. The moments are identified in advance and are divided into three different levels:

- activities; part of the journey with a specific purpose;
- touchpoints: possible aspects of the activity where the participant can receive external information from mobility provider or government;
- dimensions: aspects of the activity that might influence the experience of the participant during the activity such as availability of a seat, waiting time, perceived safety and availability of information.

These activities and touchpoints create the basis of the CJM research and help to pinpoint the opportunities for improving travel experiences of the target group. The data that is collected by the CJM research consists of a survey before the journey, questions and observational data during the journey and interviews after the journey. The survey before the journey is meant to get an overall picture of the participant focusing on digital skills, mobility behaviour and use, and experience with the journey that is part of this CJM research. The observations and questions during the journey give insights in where (touchpoint) participants experiences problems or issues and which activities or touchpoints need attention in order to improve the total journey experience of this user group. During the journey the participants are asked to score their experience for each activity using a 1 (very uncomfortable) to 10 (very comfortable) scale. The interview after the journey elaborates on this and dives into the emotions and reasons behind the experience of the participant. The overview of the DIGNITY pilots customer journey parameters are summarised in Table 2.

The outcomes differed a lot per pilot. For example, in Tilburg, elderly participants mentioned the difficulties they experience when exiting the bus or train and that it is challenging to find the right direction, especially at larger central stations. Clear signages would be helpful for them. Other participants expressed that they sometimes

Table 2. DIGNITY pilots Customer Journey mapping overview

DIGNITY pilot	# participants	Target group	Defined journey
Ancona	11	Elderly, people with disabilities	Use of local bus service and digital planning app
Barcelona	10	Low income, woman	Local bus trip to work
Flanders	7	Elderly in rural areas	Use of dial-a-bus service (it is called Belbus); a service that helps people travel by bus in less populated parts of Flanders. Reservation can be made upfront via phone or internet
Tilburg	9	Elderly, low income	Bus trip from home to train station (including preparation)

feel alarmed or uncomfortable with fellow passengers in the bus. Representatives of the low income group indicated that their preferred payment method was cash over digital methods; and that they are more often using smartphone than a computer for the digital operations. In Flanders pilot, participants appreciated possibility to use the phone for making reservations of the dial-a-bus service. Especially for people with low digital capabilities this service is very successful. In Ancona, almost all participants mentioned it was easy to find the ticket validator machine inside the bus. But when they needed help, only 5 out of 11 participants perceived the bus drivers as friendly and helpful to assists the process.

Figure 4 illustrates the scores that participants gave for each of the journey activities in the CJM research in Ancona. Each line represents a participant and each number corresponds to the feeling participants had during that activity: 1 means very uncomfortable and 10 is very comfortable. Some of the scores of participants are the same and therefore are not visible in this graph. The activity with the lowest scores has the potential to be improved which will be beneficial for the overall experience of the journey.

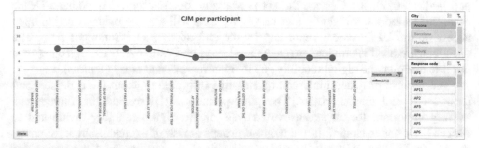

Fig. 4. Ancona pilot CJM participant scores.

The information gathered with the CJM method provided a detailed and clear insight on the experience of the participants. It enriches the quantitative data that is collected in the self-assessment tool and shows specific examples of how the participants experience the journey. Conclusions need to be carefully drafted and interpreted, since the method involves the experiences of a limited group of people. Overall, pilots expressed that using this method gave them new information about the experience of the product/service. Both the product/service developers and public authorities appreciated it for the insights on the usability of product or service by a specific user group, allowing to get information for potential improvements in order to increase the user experiences of the user group.

3.3   Large Scale Surveys

The understanding of the digital mobility gap can also be informed by large scale surveys. As part of the DIGNITY project, a questionnaire was developed to support this, examining a range of factors that affect the use of digital mobility systems. It was based on an earlier survey conducted in the UK in 2019 (Goodman-Deane et al. 2021) and covers technology access, technology use, limitations in travel, attitudes towards technology and basic digital interface competence. Most of these were assessed using multiple-choice self-report questions. The exception was digital interface competence which was measured using a simplified paper prototyping method. Participants were shown paper mock-ups of smartphone interfaces and indicated on the mock-ups what they would do next to achieve eight simple tasks. The questionnaire was initially developed in English and then translated into the local survey languages by professional translators.

Surveys were conducted using this questionnaire in five countries or regions, including four related to the DIGNITY pilot areas (the Barcelona Metropolitan Area, the Netherlands, Flanders and Italy) plus Germany. All of the questionnaires were administered in face-to-face interviews, to enable the inclusion of people without Internet access and obtain a better picture of the digital mobility gap. The surveys were conducted at different times in 2020 and 2021, as was possible under local COVID-19 restrictions. All surveys were conducted in a manner compliant with these restrictions, maintaining social distancing and wearing face coverings as appropriate. Quota sampling, area sampling and stratified sampling methods were used in the different surveys. Ethical approval for the surveys was obtained from the University of Cambridge Engineering Department ethics committee. More information on the surveys is available in (Goodman-Deane and Waller 2022), the German dataset is available open access at (Goodman-Deane et al. 2022) and the remaining four survey datasets will also be made available open access on the UPCommons repository by the end of 2022 (UPCommons n.d.). The questionnaire itself is provided in (Goodman-Deane and Waller 2022) so that others can use it to examine the digital mobility gap in other regions and areas (Table 3).

The surveys provided important quantitative information about the end-users and their needs and characteristics. The pilot partners all described the survey data as very important for an exhaustive analysis of the digital mobility divide and to deepen their understanding of the targeted groups at risk of exclusion. In some cases, the DIGNITY surveys were the only obtainable data sources about the topics and vulnerable-to-exclusion groups of interest to the pilot.

In general, it is important to try to achieve as representative a sample as possible and compare the survey demographics with those in the general population to help understand sample biases. The recruitment and sampling in the DIGNITY surveys were particularly

Table 3. Summary of DIGNITY survey data.

DIGNITY pilot	Survey location	Date	# parti-cipants	Weighting	Gender distribution (unweighted sample)	Age distribution (unweighted sample)
Ancona	Italy	Nov 2020	1002	By age, gender and region	Male 49% Female 51%	Age 16–39 26% Age 40–64 50% Age 65+ 24%
Barcelona	Barcelona Metrpolitan Area	Nov-Dec 2020	601	None	Male 48% Female 52%	Age 16–39 35% Age 40–64 42% Age 65+ 22%
Flanders	Flanders	June-Sep 2021	418	By age, gender and region	Male 49% Female 51%	Age 16–39 42% Age 40–64 36% Age 65+ 22%
Tilburg	The Netherlands	Sep 2020, July-Sep 2021, Nov 2021	423	By age and gender	Male 49% Female 51%	Age 16–39 37% Age 40–64 37% Age 65+ 25%
N/A	Germany	July – Sep 2020	1010	By age, gender and region	Male 48% Female 52%	Age 16–39 33% Age 40–64 41% Age 65+ 20%

affected by the COVID-19 pandemic. In particular, some potential participants may have been wary about taking part in a face-to-face interview due to the risks of infection. This is likely to disproportionately affect older people and those with underlying health conditions or disabilities. Both of these groups have lower levels of technology use and competence (Goodman-Deane et al. 2022). In addition, people who are less interested in technology may have been more reluctant to take part in a survey about technology. As a result, the surveys may underestimate levels of digital exclusion.

Some participants had difficulties understanding some of the technological concepts and experiences mentioned in the questionnaire. This could hamper statistical analysis as it is difficult to determine whether a response of "don't know/prefer not to answer" can be attributed to a lack of knowledge/confidence or to the interviewee being tired or confused. In the surveys, this was addressed by the interviewers working to encourage the participants to collaborate and to keep their attention, so that the interviews could be completed successfully. Some extra explanation of technological aspects could also be added to the questionnaire.

The study used a paper prototyping method for assessing basic digital interface competence. This is in contrast to the self-report methods commonly used in large scale surveys and provides a more reliable and direct insight into participants' technology competence. However, it is less reliable than tests carried out on live interfaces, where participants can explore the interface and try out different actions. This limitation was mitigated by selecting straightforward tasks in which success was largely dependent on a single tap of something currently visible on the screen. Another issue is the sampling variation between countries which makes cross-country comparison difficult. For example, Germany and Italy had large samples that attempted to be population-representative while the other surveys had smaller, less reliable samples.

The results from the surveys indicate that substantial numbers of people in all the surveyed countries lack access to or do not use digital technology. For example, between 6.0% (in the Netherlands) and 19.7% (in Italy) of the sample had not used a smartphone in the previous 3 months (see Fig. 4). Digital mobility services requiring the use of a smartphone app are likely to be particularly exclusionary because the user needs to know how to install an app as well as use a smartphone. Furthermore, many people have low levels of basic digital interface competence, ranging from 18% in Flanders to 33% in Italy. These figures indicate that there are large numbers of people who use technology but are still likely to struggle with several aspects of a basic smartphone interface (Fig. 5).

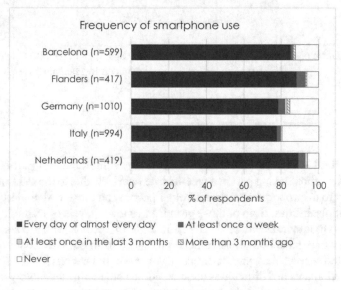

Fig. 5. Frequency of smartphone use.

The use of digital mobility services was low. Between 35% in Flanders and 87% in Italy had never used *any* of the services examined in the survey (car sharing, carpooling, digital taxi services, on-street bike hire, scooter/motorbike hire and mobile phone parking payment). However, it should be noted that differences between regions may be due in part to different levels of roll-out and availability of these services. The findings indicate that there is a long way to go before these services truly become mainstream, and it is important that service providers do not assume that users have familiarity with how they work.

There were also high levels of travel limitations: those reporting being very limited in their regular travel in the region ranged from 27% in the Netherlands to 45% in Germany. The main reasons for these limitations varied between regions. Common reasons were the cost of travel, limited availability of transport services and transport infrastructure and safety concerns. In addition, substantial numbers reported limitations because digital skills were needed to plan the travel or use the transport.

3.4 Focus Groups

A focus group brings the DIGNITY framing methodology to its conclusion. Building on a long tradition of social science research, focus groups are organised to facilitate discussions within a carefully selected small group of people. Interaction between the participants is the key distinctive characteristic of a focus group. A focus group provides insights in group dynamics, on how people form their opinion and help to better understand the perspective of the group that is being studied.

Within DIGNITY framing phase, it allows to take into account the perspectives of the vulnerable-to-exclusion end-users of the digital mobility products and services. This specific knowledge provides an added value to other data collection methods deployed in other DIGNITY framing methodology steps (Bracke et al. 2021). With this method, there is no ambition to collect a lot of new data. It enables the collection of in-depth, qualitative data on a micro level, with the goal to contextualise and better understand the already collected data. For this reason, the focus group is an ideal method to discuss and validate the results of the previous steps in the DIGNITY framing methodology.

Each DIGNITY pilot city or region were responsible for organisation and moderation of one focus group, with the number of participants ranging from 7 in Flanders to 21 in Ancona. Since the target group were people who are vulnerable to digital exclusion, a face-to-face setting was aimed for. Table 4 gives an overview of the focus group organised in each pilot. In all four focus groups elderly were represented and a gender balance was achieved. Recruiting participants from vulnerable-to-exclusion groups for a live discussion proved to be difficult, especially during the COVID-19 pandemic. Still, three out of four pilots managed to organise a (partly) face-to-face focus group, while only Ancona had to switch to an online alternative.

Given that the focus group builds on the results of the customer journey mapping and the survey mainly, the content was not fixed in advance. The topics and specific questions depend on the data and insights from these previous steps. Table 4 therefore also lists the topics discussed in each DIGNITY pilot city or region. In line with the focus of DIGNITY, two topics were discussed in all pilots: how the participants experienced limitations in their daily travel due to digital reasons; and if they think there is too

Table 4. Overview of the focus group organised in each pilot.

DIGNITY pilot	Format	# partici-pants	Target group	Topics discussed
Ancona	Digital	21	None specifically targeted, elderly, migrants and visually disabled were present	Use and trust in local mobility app, asking others (bus drivers, fellow passengers) for help during a trip, information at bus stops, personal safety
Barcelona	Face-to-face	10	Low income, migrants, elderly	Information needed to plan a trip and how to look it up, asking social network for help when digitally planning a trip, financial limitations, information at bus stops, personal safety
Flanders	Hybrid	7	Elderly in rural areas, disabled	Information needed to plan a trip, availability of travel services in the neighbourhood, reasons to opt non-digital solutions, attitude towards "Belbus" (demand responsive transport)
Tilburg	Face-to-face	8	Elderly, physically disabled	How to prepare for a trip, asking social network for help when digitally planning a trip, coping with unforeseen circumstances during a trip, use of a chipcard, financial limitations

much focus on digital solutions in mobility and whether a balance between digital and non-digital services should be aimed for.

To structure the method, a template with specific questions for each pilot was prepared. This template was completed by the pilots and further analysed by the DIGNITY research partners. A more collaborative way of designing the template with questions and the analysis might be recommended for future use, e.g. building on the pilots understanding of the local situation and the research partners knowledge could lead to a more applicable and relevant topic list for each city or region. Being responsible for the organisation and moderation, pilot representatives were the only project partners who attended the focus groups. Therefore, a more direct involvement from them in the analysis, which was now done exclusively by the research partners, can also improve the analysis part of this method.

The focus groups were very well perceived by the pilot partners as well as from the representatives from the vulnerable to exclusion groups, who greatly appreciated the chance to be heard on the topic. Especially the face-to-face organisation, which enabled direct contact with the target group of the project, was very much appreciated. This showed the value of face-to-face, live research, even during the pandemic. Given that only one focus group in each pilot already delivered very interesting and valuable results, it is recommended to organise more than one focus group, as was the case now.

The focus groups provided extra insights into the experience of vulnerable-to-exclusion groups with digital mobility. Despite the questions being specific to the local situation, there are some commonalities in the results. These help to better understand the digital gap in mobility. First, the representatives of the vulnerable-to-exclusion groups, especially the elderly, confirmed that digitalisation might hinder them in their daily mobility. According to the Flanders focus group, working with digital services or products is often too complex, while the requirement of digital ways to buy a ticket or find information might be a reason to postpone or even cancel a trip. Participants in Ancona were worried they might do something wrong and cause the digital service to break. They therefore advocated to simplify procedures that users have to go through when using digital services. Next, in all focus groups, participants stressed the importance of personal contact for help. Both in preparing a trip as during a trip, participants indicated they often have to ask others for help, because they lack the necessary digital skills. During a trip, this mainly means turning to fellow passengers or staff, if they are present at all. In Barcelona, this is thought of as a last resort, since bus drivers are often not perceived as helpful or they don't seem to know the answer. If people need help looking up information or buying a ticket in preparation of a trip on the other hand, most participants need to rely on their social network. In Tilburg, elderly participants most often turn to children and grandchildren if digital actions are necessary. They indicated that this is not always easy, as they do not want to disturb their social network. Without a proper social network assistance for digital actions becomes more difficult, which might lead to further social isolation. Lastly, in several focus groups there was also optimism about the possibilities that digitalisation might bring about. In Flanders, for example, it was mentioned that digitalisation could add efficiency to the transport system and lead to a better integration of services. But for this to properly work and be inclusive, it is essential to involve users from all parts of society in the design of mobility products and services. Participants in Ancona mentioned another possible solution by organising easy, accessible trainings so the most vulnerable users can learn how to use digital mobility solutions.

4 Discussion

The implementation of the DIGNITY framing methodology has been concluded by all the pilots at the end of 2021. Methodological soundness and applicability of these methods were further evaluated and validated through the DIGNITY validation approach (Lazzarini B. 2022), which provided detailed feedback on each method from the variety of the involved stakeholders. The objective was to assess the usefulness and effectiveness of individual methods deployed as well as an added value of the overall framing phase to different local/regional context. The evaluation of the framing phase has been carried out using a set of evaluation criteria, further detailed with indicators. Evaluation criteria were: effectiveness; efficiency and resources; participation and collaboration; expectations & social learning/capabilities acquired'; relationship with other DIGNITY tasks.

Overall, all the methods proposed within the framing phase were considered useful by pilots. Figures 6 and 7 illustrate it: pilot partners placed themselves on the right upper quadrant of the scheme, which describes their overall satisfaction with the framing phase. For example, pilot partners described without exception that the activities of the framing phase were very important for raising awareness of the problem of the digital divide related to mobility. It also enabled the collection of essential information for the implementation of local initiatives (Lazzarini B. 2022). Data collected through the variety of methodologies, allowed to contextualise the digital gap in a specific geographical context and to better identify the vulnerable to exclusion groups for the further policy focus.

Improvement can be achieved in the integration of the different insights/results provided by framing methods, considering the fact that the information collected through the different methodologies is quite diverse. Next, as a result of the framing phase it became evident, that there is currently a lack of public data focusing on the digital gap in mobility of vulnerable-to-exclusion groups and the need to a systematisation of a set of standard data, ideally by public administration/entities in order to support decision making.

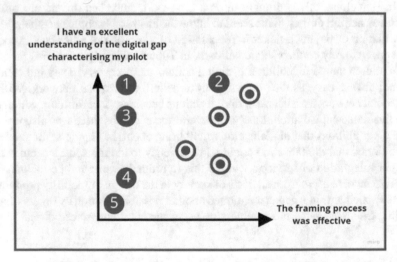

Fig. 6. Evaluation of the framing phase by DIGNITY pilots.

Fig. 7. Evaluation of the framing phase by DIGNITY pilots.

DIGNITY framing phase proposed a portfolio of methods to assist the cities and regions in identification of the impacts that the mobility eco-system digitalisation has on different population groups. This process allows to prioritise the vulnerable-to exclusion groups to which the most attention needs to be paid on the local level within a digital mobility transition process; to identify digital mobility products and services inclusiveness of which can be improved and to distinguish the gaps in institutional, organisational and regulatory structures within public authorities allowing to build inclusive mobility eco-systems.

Building up on these results, the next steps within DIGNITY approach allows to act on those shortcomings. The bridging phase includes:

- The scenario building approach aims to analyse possible developments in the future and to present them coherently; focusing on the gaps identified at the macro levels.
- The iterative process of inclusive design wheel, offering a structured method for generating solutions to challenges, with an emphasis on creating solutions that are usable by as many people as reasonably possible, this way aiming at the gaps identified at micro and macro levels.

The outputs of the bridging phase are used to develop a robust regulatory framework and policy action plans as well as to develop more inclusive mobility products and services, addressing the needs of the variety of the users. Pilots confirmed that overall set of methods developed in DIGNITY have improved the understanding of the digital gap, at different levels and allowed to move forward in the development of the inclusive mobility eco-system.

5 Conclusions

Digitalisation is one of the current trends in society, that facilitates the connectivity between people, businesses, regions, and countries. Location and distance are no longer a barrier to meeting and exchanging information (Hoeke et al. 2020). Digitalisation of the transport sector follows a high speed path, changing the ways people access information about transport services and products, as well as changing mobility patterns and the use of some transport options. However, not everyone benefits from these digital developments, bringing specific population groups at risk of being excluded from some mobility products or services and creating a risk of the increasing social isolation. There is no general solution that exists, as this gap manifests itself in different ways taking into account the local situations. Public authorities feel the urge of assessing potential scope of the problem and realise that specific vulnerable to exclusion groups might require growing attention. The framing methodology proposed in DIGNITY, builds on variety of individual methods that allow public authorities to develop an in-depth view on the scope, size and urgency of the problem. Methods presented in this chapter are proven to be useful as standalone activities, however their maximum impact is achieved in their joint implementation.

References

Banister, D.: Transport for all. Transp. Rev. **39**(3), 289–292 (2019). https://doi.org/10.1080/014 41647.2019.1582905

Bracke, A., Delespaul, S., van Apeldoorn, N.: DIGNITY, Framing the digital gap in mobility on a local level (D.3.1.) (2021)

Goodman-Deane, J., Bradley, M., Clarkson, P.J.: Relating age, digital interface competence and exclusion. Gerontechnology **20**(2) (2021). https://doi.org/10.4017/gt.2021.20.2.24-468.11

Goodman-Deane, J., Waller, S.: D1.2 Benchmark of factors affecting use of digital products and surveys across Europe. DIGNITY project deliverable D1.2 (2022a). https://www.dignity-pro ject.eu/wp-content/uploads/2022/03/Dignity-D1.2-final.pdf. Accessed Apr 2022

Goodman-Deane, J., Kluge, J., Waller, S., Roca Bosch, E.: DIGNITY digital exclusion dataset - Germany . 1 v. Universitat Politècnica de Catalunya (2022b). https://doi.org/10.5821/data-2117-367811-1

Goodman-Deane, J., et al.: Towards inclusive digital mobility services: a population perspective. Interact. Comput. **33**(4), 426–441 (2022)

Hoeke, L., Noteborn, C., Goncalves, M., Nesterova, N.: DIGNITY, Literature review effects of digitalization in mobility in society (D.1.1) (2020)

Lazzarini, B., Carmona, N., Roca, E.: DIGNITY, DIGNITY framework validation report (D.4.3) (2022). Forthcoming

Nesterova, N., Goodman-Deane, J., Bradley, M.: DIGNITY, Interactive catalogue with good examples of mobility products for vulnerable groups, including an overview of success factors for mobility products/services to meet user needs of vulnerable groups (D 1.3) (2021)

UPCommons (undated): https://upcommons.upc.edu/. Accessed Apr 2022

Creating a More Inclusive and Accessible Digital Transport System: Developing the INDIMO Inclusive Service Evaluation Tool

Hannes Delaere[1]([⊠]), Samyajit Basu[1], and Imre Keseru[2]

[1] Mobilise Mobility and Logistics Research Group, Department Business Technology and Operations (BUTO), Vrije Universiteit Brussel, Brussels, Belgium
Hannes.Delaere@vub.be
[2] Mobilise Mobility and Logistics Research Group, Vrije Universiteit Brussel, Brussel, Belgium

Abstract. The introduction of smart technologies in mobility has created a vast landscape of possibilities and options, but at the same time they have also created uneven impacts across society. It is, therefore, the goal of this paper to introduce the online Service and Policy Evaluation Tool (SPET) for evaluating the accessibility and inclusivity of digital mobility and delivery services. The tool shall enable policy makers to design strategies necessary for all citizens to fully benefit from the digital mobility system (e.g. social and educational strategies, new regulations, etc.) and identify strategies to avoid digital exclusion in terms of social and spatial aspects. Structurally, the tool is built on the capabilities approach, in combination with the principles of universal design, and co-creation was used for the development of the tool contents. The recommendations from the SPET will assist policy makers, developers, operators and other parties to provide promised benefits of digital services to all sections of the society, especially to people vulnerable to exclusion.

1 Introduction

In order to participate in social or other activities, a person needs to navigate the environment (Vecchio and Martens 2021). The transport environment used to be a purely physical one, but since the introduction of the internet and especially the smartphone, there is an increasing need for digital skills to navigate this new digital transport environment (Vaidian et al. 2019; Velaga et al. 2012). This has, however, proven difficult for many groups in our society, resulting in groups of people that are, more than others, vulnerable to be excluded from participating in social or other activities (Groth 2019; Loos et al. 2020; Pangbourne et al. 2020). Moreover, the proportion of people having access to mobile internet access was still only 74% in 2019, and this form of internet connection is most relevant for using digital services while on the road. The percentage of Europeans who recently (up to 3 months before the Eurostat data collection in 2020) used an internet connection to order online transport service, from an enterprise or private person, is only 9%. For older people, who are often already struggling to fulfil their mobile needs, this is only 4% (Eurostat 2020), indicating the need for improved accessibility and inclusivity of digital transport services.

I. Keseru and A. Randhahn (Eds.): *Towards User-Centric Transport in Europe 3*, LNMOB, pp. 254–274, 2023.
https://doi.org/10.1007/978-3-031-26155-8_15

Research shows that several characteristics, such as education, income, gender, age, migration background etc. have a significant impact on a person's access to digital transport services (Durand et al. 2022; Estacio et al. 2019; Gorski 2005; van Dijk 2006). These traits are usually combined, e.g. a person with a migration background, who did not receive extensive education, will often earn less and has fewer digital skills, resulting in difficult barriers for her/him to use a digital mobility service (Durand et al. 2022; Sathyan et al. 2022). The lack of access to digital transport solutions is a result of the combination of limiting socio-economic factors and already existing transport disadvantages. Durand et al. (2022) argue that increased exclusion from the digital transport system is already developing within the same groups that currently experience transport disadvantages and a higher degree of social exclusion. The further digitalisation of the transport network is likely to create new lines of transport and social inequality, as well as enforce existing ones. Although some studies have already linked digital exclusion and transport disadvantages, not much empirical research is available yet.

The capabilities approach has been proposed as a possible evaluation approach to appraise the contribution of transport projects and services to wellbeing and freedom to access opportunities (Vecchio and Martens 2021). If we apply the concept of the capabilities approach to digital mobility services, in order to use a service, a person needs to have the ability to utilise mobility resources (e.g. public or private transport). This ability is influenced by knowledge, skills, confidence, physical and mental ability to access and navigate on a digital interface. To increase this ability, it is necessary to provide digital mobility services that require fewer skills and other resources, and which are accessible and inclusive towards people of all abilities.

In this paper, based on definitions by Lucas (2012) and Schwanen et al. (2015), we consider inclusive digital mobility to be a very variable concept, influencing and influenced by any party that is involved, creating a combination of subjective and objective combination of expectations, needs and barriers that need addressing before a digital transport service can be considered to be inclusive. These expectations, needs and barriers can be addressed from different standpoints, given the multitude of stakeholders of the digital transport network. For a definition on accessibility we adopt the definition developed by the European commission which states that accessible digital mobility has to comply with the following two aspects: "Provision of appropriate and sufficient information for the passenger to plan and carry out the journey, and to deal with unexpected disruptions. Provision of the information in the format and via the channel suited to the passenger, especially considering those with visual, hearing, learning and cognitive difficulties" (European Commission et al. 2020, p. 6).

In combination with the capabilities approach, universal design can be a useful concept to create a digital transport system that is accessible and inclusive. The principles of universal design focus on developing a spatial environment that is physically accessible to all (Mace et al. 1998). This means that the environment is developed to fit the skills of all people, including those with a physical disability. When applied to the digital transport system, this results in the development of digital services that need to comply with the needs of people with the lowest digital skills. Creating a digital transport system that answers to those needs requires a co-creative and inclusive approach, with input from those people that are involved in the development process (developers, operators, policymakers etc.) of digital mobility services and of especially those groups that are currently excluded from these services due to a lack of skills.

In this paper, we introduce the INDIMO Service and Policy Evaluation Tool or SPET, the first tool of its kind that evaluates the inclusivity and accessibility of digital mobility and delivery services. This tool aims to bridge the gap between the abilities of persons vulnerable to exclusion and the requirements of emerging digital mobility and delivery services by providing an online self-assessment tool for evaluating the inclusivity and accessibility characteristics of an existing or new, yet to be introduced service. The tools also provides recommendations on how to improve specific features of an evaluated service so that the accessibility and inclusiveness of the same is improved.

The following research questions need to be addressed for developing a Service and Policy evaluation tool that is efficient and useful.

– How can we facilitate the evaluation of the inclusiveness and accessibility of digital mobility services?

  • Which topics are significant for the development of a digitally inclusive and accessible tool?
  • How are the different key topics included in the tool evaluated and scored?

To answer these questions, we first introduce three key concepts: the capabilities approach, the principles for universal design, which were used as the basis for the tool and the co-creation method which was used to develop and test the tool. Secondly, the conceptual framework is presented, then the methodology of developing the tool is discussed and finally, in the output and structure section, we explain the development of the tool step-by-step, how we evaluated the questions, how the weights were allocated to different topics and how the performance scores are calculated. In the conclusions, we propose additional functions and services that can be included in the SPET, as well as future research on the content and use of the SPET.

2   Literature Review

2.1   Capabilities Approach

Current transport planning and policy are mostly focused on the transport system itself, without actually focusing on those using the system, resulting in the idea that a decent or good working system is enough to provide transport for everyone, indirectly indicating that all people have the possibility to participate in activities (Brown et al. 2009). It cannot be denied that this approach has provided an ever increasing accessibility to a significant part of the population, however, now it is obvious that this does not mean this method has proved to be sufficient for everyone (Lucas 2012). The ever-increasing digitalization of the transport system, and consequentially the increased complexity of the same demands a different approach.

The need for a new approach emerged in the 1990s, with the introduction of the capabilities approach, a theory developed by Amartya Sen and Martha Nussbaum (Nussbaum 2000, 2011; Nussbaum et al. 1993; Sen 1985, 2001, 2009). Various definitions and descriptions have been adopted, for this paper, we adopt the following definition as proposed by Sen (1995, p.1)" A person's capability to achieve functionings that he or she has reason to value provides a general approach to the evaluation of social arrangements, and this yields a particular way of viewing the assessment of equality and inequality". The capabilities approach was also promoted as a methodology for the appraisal of transport systems, with its foundation in the contribution it provides to a persons' opportunities and wellbeing, which is a basis for consistent evaluative approaches to influence transport planning and policies (Alkire 2003; Vecchio and Martens 2021).

Besides its ability for evaluating transport systems, the capability approach inherently promotes accessible transport systems and thus has a positive impact on the groups in society that are more poorly served by the current transport system than others (Lucas 2012; Martens 2017). Furthermore, it is especially useful when it is used for evaluating a diverse set of people, each with their own capabilities and constraints, keeping in mind the distribution of mobility resources and how these are differently available and used by different people (Vecchio and Martens 2021).

The capabilities approach, especially in relation to mobility has been approached in multiple ways, from a very broad interpretation: 'the ability to be mobile' (Beyazit 2011), from a physical, social and financial point of view, to 'being able to use public transport' (Ryan et al. 2015). Another possible approach was introduced by interpreting accessibility as a capability, rather than just being mobile. This interpretation focused on the participation in society, for which a person needs the be mobile to a degree (Martens 2017).

Lastly, another important difference in interpretation and use of the capabilities approach is combining it with a top-down or bottom-up approach. For the creation of a service or product for people vulnerable to exclusion, the bottom-up approach is preferred as this examines how each person attributes different values to an activity and how this results in participation in activities due to the accessibility provided through the transport system (Vecchio and Martens 2021). In other words, the bottom-up approach includes those people in the development process for whom the product or service is meant. Contrary to the top-down approach, where users are not involved in the development process. A main disadvantage of the top-down approach in this case is the lack of knowledge about barriers and needs that are experienced by people vulnerable to exclusion, as well as other stakeholders of the digital transport network. Therefore, in this paper, we consider the bottom-up approach to be the most suited approach when researching and working with citizens vulnerable to exclusion. In combination with the capabilities approach of Randal et al. (2020), this results in a policy and service evaluation tool that was developed with input from users, developers, operators etc. so that their capabilities, requirements and needs are integrated in the tool.

Universal Design Principles in Digital Service Design

Universal design (UD), a concept first mentioned and used by Ronald Mace, an architect who worked on social inclusion of people with disabilities, is described as "the design of products and environments to be usable by all people, to the greatest extent possible, without the need for adaptation or specialized design" (Aarhaug 2019, p. 2). This resulted in a first, comprehensive approach to develop a more inclusive world. Earlier attempts resulted in a segregated approach, one infrastructure for those without 'disabilities' and one for those who needed an adapted approach. These attempts usually were considered to be ugly and required additional investment (Mace et al. 1998). Throughout the years, the way UD was approached has evolved and a vision emerged that UD promotes a new approach to design that celebrates diversity and provides equal opportunity of access to mobility and services (Audirac 2008). In the EU, as part of the General accessibility act (2019), the concept of UD is defined and integrated within the legislation for the development of products and services (European Commission 2019). A more recent practical evolution is the implementation of UD in digital services, with Begnum and Bue (2018) concluding that there is still a significant lack of awareness of UD among designers and other stakeholders.

The widespread emergence of services, especially digital services, has resulted in the fast growth of service design, focusing on the holistic experience of users, with the goal to make services user friendly, easy to use and more intuitive (Polaine et al. 2013; Scott et al. 2016). According to Begnum and Bue (2021) a widely accepted definition of an inclusive digital service is still not available, so, in this paper, we adopt the working definition as described in their work: "A service is universally designed when its customer journey is usable to all people (to the greatest extent possible and without the need for adaptation or specialized design), by selecting suitable touchpoints" (Begnum and Bue 2021, p. 22). The design has a significant impact on the value creation of the digital services (Law et al. 2008) and consequently on society as well (Kuk and Janssen 2013), with digitalized services dominating society at an ever increasing pace (Newman 2020). The impact service design has, is significant, and even though there is an idea on the relevant concepts linked to universal service design, awareness about this method is overall lacking (Begnum and Bue 2021; Delaere et al. 2020).

The introduction of universal design, or inclusive design, which are often used as synonyms (Goodman-Deane et al. 2010; Clarkson and Coleman 2015) in digital services has resulted in the development and design of more inclusive digital services. Adopting universal design principles when developing services is not necessary for most users, but it does provide the opportunity for vulnerable to exclusion people to make use of the services as well, resulting in a more accessible and inclusive service for all users. In this regard, one set of regulations that has proven impactful are the Web Content Accessibility Guidelines, which are guidelines defining how to make Web content more accessible to people with disabilities. Accessibility involves a wide range of disabilities, including visual, auditory, physical, speech, cognitive, language, learning, and neurological disabilities (W3C 2008). Within the European digital landscape, the Web Content Accessibility Guidelines 2.0 (WCAG 2.0) are mandatory for National agencies, and their contractors (European Union Agency for Fundamental Rights 2014). Additional guidelines are available in the WCAG 3.0, which have significant overlap with WCAG 2.0, but it introduces an alternative to previous versions (W3C 2021). Although WCAG has resulted in an increase of the accessible character of web based applications and platforms, since social and spatial inclusion is not really integrated into the definition of the WCAG, it does not provide an answer or guideline to all barriers of digital services (Begnum et al. 2018). Tools have been developed to assess the inclusive character of projects, programs, organisations and companies, but only a few have been created for the evaluation of digital services (Department for Digital, Culture, Media and Sport 2017), with none focusing on digital mobility services.

The implementation of UD principles in legislation shows an effort for a more generally accessible design and use of services and products. Nevertheless, there is still a lack of knowledge among developers, designers regarding UD in service design. Moreover, inclusivity of services and products is often only a sidenote in the company goals or not mentioned at all (Delaere et al. 2020). This is where the SPET can fill an existing gap, it can provide assistance to evaluate the digital inclusivity of digital mobility services and can provide guidelines to design, develop and implement accessible, inclusive digital mobility services.

2.2   Co-creation as an Approach to Develop Tools

Pappers et al. (2020) stated that the use of co-creation, as a form of public participation has become more present in multiple industries, especially within health and education. In transport development and policy, co-creation has only recently become more of a standard, advocated by the European commission with projects like those within HORIZON2020.

Developing digital mobility services that are accessible and inclusive require user-centred development, and applies a bottom-up approach, meaning that the users are involved in the development process and have a significant impact on the final output. One way of involving users in the design of a service is using 'co-creation', that can be interpreted as a more intense, further reaching form of customization, involving "collaboration with customers for the purpose of innovation" (Kristensson et al. 2008, p. 47). In a co-creation process the focus shifts from the firm or company to the users, who are significantly involved in the development process addressing user- specific needs (Chathoth et al. 2013).

The first stakeholder group, mainly the users, are mostly the sole focus of inclusivity measures, as these are the people that might potentially be excluded. But, for the development of a digital service, it is not sufficient only to keep in mind the users, as this can result in demands that cannot be fulfilled by developers, operators or policy makers. Therefore, it is important that both sides have sufficient input in the development of the tool. This will eventually result in a service that is inclusive and accessible for all users, but will also make sure that all inclusion related changes and measures are feasible for the developers, operators and policy makers to implement as well.

Conceptual Framework
The Service and Policy Evaluation Tool, that we developed, addresses the gap between the provided capabilities of potential users, and the required capabilities of a digital mobility or delivery service based on the capabilities model of Randal et al.(2020). It is intended to align the set of requirements posed by a digital service to match the capabilities of potential users to facilitate their participation in society. The tool aims to help the key stakeholders that have an influence on the development of digital applications and services, i.e. developers, operators and policy makers, to make digital mobility services universally accessible and inclusive.

The goal of the SPET is to intervene in matching both sides of the capabilities model (Fig. 1). Provided capabilities (e.g. digital skills), have to match the required capabilities (constraints) in order for a person to participate in a specific activity (e.g. visiting a family member), which are influenced by the actual activity (e.g. using a ride-sharing service) and the environment, i.e. digital and physical context in which the activity will take place (e.g. the smartphone application that is needed to book a ridesharing service as well as being able to find a physical meeting point to get to the vehicle)(Vecchio and Martens 2021). Applying universal design to the service and the environment could help to design a service that accommodates the capabilities of as many people as possible rather than designing specialised services addressing capability limitations. In order to facilitate universal design, developers need to take it into account when designing the service and its interface; operators need to consider it when they operate a digital mobility service; and policy makers need to create guidelines and regulations that incentivise developers and operators to comply with accessibility and inclusivity requirements. If the Universal Design principles are considered, the activity and the environment can be designed in a way that accommodates the requirements of the users in a broad sense and the capability gap disappears or it is at least decreased. In this way, a user would be able to book a ridesharing service through a smartphone app, communicate with the driver and find the meeting point and board the vehicle in order to reach her/his destination irrespective of her/his level of digital skills.

The SPET has both an assessment and a steering role in this process. On the one hand, it would allow policy makers, as well as developers and operators to assess to what extent a service complies with minimum and recommended accessibility and inclusiveness standards; but on the other hand, it would also steer service design by giving recommendations on applying universal design to improve specific features of services and applications.

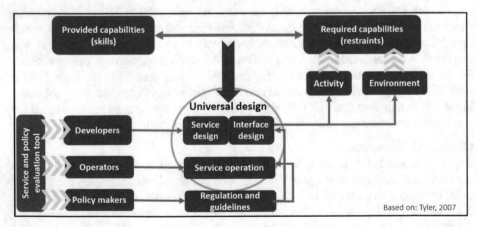

Fig. 1. Conceptual framework of SPET

3 Methodology of the Development of the SPET

The SPET was developed through a co-creation process involving the two main stake-holder groups: the citizens, i.e. the users and non-users of digital mobility services and the stakeholders that are involved in the development, implementation and operation of the digital mobility services (developers, operators and policy makers). The creation of the SPET was a multi-step process, presented in Fig. 2.

In the first step of the development the requirements of the stakeholders towards the tool were identified. We carried out 10 case studies of digital mobility and delivery services to assess how accessibility and inclusion were considered in their development. As part of the case studies, 18 interviews with operators, developers and policy makers provided information about the drivers and barriers they experience during the development, implementation and operation of an inclusive digital mobility service (Delaere et al. 2020). Then a co-creation workshop was organised where 36 experts discussed and elaborated on the drivers and barriers that were collected from the interviews. This provided us with a more extensive understanding of the barriers, as well as the differences between different kinds of stakeholders.

In the second step, we defined the set of capabilities of users and non-users (skills) and the requirements of digital mobility and delivery services (constraints) through a comprehensive qualitative research process. 70 interviews were conducted with users and non-users of the digital transport system with a focus on the people vulnerable to exclusion. Additionally, 25 interviews were conducted with stakeholders representing people vulnerable to exclusion as part of 10 user case studies. In these interviews we investigated the required capabilities to use a digital mobility service and the provided capabilities by people vulnerable to exclusion (Ciommo et al. 2020a, b; Vanobberghen et al. 2020).

In the next step, guidelines were developed for applying universal design in the development of digital mobility and delivery services and applications. The first set of guidelines were collected in the Universal Design Manual – (UDM)(Ciommo et al. 2020a, b); secondly, there are the guidelines for universal language interface icons and accessible interfaces (UIL)(Hueting et al. 2020), and the last set of guidelines are those about cybersecurity, privacy assessment and data protection (CSG) (Capaccioli et al. 2020). These documents provided the set of minimum requirements and recommendations to be used in the SPET. These are represented as a set of themes and topics (see in the next section), as well as the questions that are included in the SPET.

A co-creation workshop with 28 mobility experts was organised during which the topics and questions were reviewed, discussed and changes were suggested. To finalize the development of the questions, the same process was repeated a second time with 19 developers, operators and researchers.

In the fourth step, weights and scores were developed for the questions in the SPET since during the co-creation workshops, it became clear that the mobility and inclusivity experts did not consider each topic to be equally important. Thus, a weighting exercise was organised to find out which topics they considered to be relatively more important compared to others.

In the fifth step, an online web version to facilitate the accessibility and inclusivity assessment, was developed, tested and evaluated.

In June 2022, a first test for the SPET was organised with 24 experts comprising of researchers, developers and policy makers from cities across Europe. During this test the participants used the SPET to evaluate several digital mobility and delivery services: Cambio[1] (carsharing), Bpost[2] (eletronic parcel locker), Uber Eats[3] (food delivery), BlaBlaCar[4] (ridesharing) and Citymapper[5] (multimodal routeplanner). While completing the evaluation of one of these services the participants answered a survey about the clarity and understandability of the questions, answers, definitions and results in the SPET.

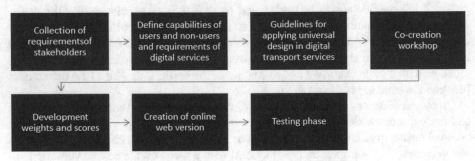

Fig. 2. Stepwise creation process of SPET

4   Structure and Output of SPET

This section has the following structure: first the different types of evaluations are explained, focusing on the type of service and the evaluator, secondly the themes, topics and related questions are defined, thirdly we talk about the structure of the questions, a fourth section focuses on the weighting of the topics, followed by an explanation on how the scoring takes place and finally how the output of the tool is produced.

4.1   Types of Evaluation in the SPET

Before evaluating the digital inclusivity and accessibility of a mobility service, the evaluator will have to select what kind of service will be evaluated and which type of stakeholder (developer, designer, operator or policy maker) she or he is. Based on these choices, a specific set of topics and questions is presented to the evaluator.

---

[1] www.cambio.be.

[2] www.bpost.be.

[3] www.ubereats.com.

[4] www.nl.blablacar.be.

[5] www.citymapper.com.

The services currently available for evaluation are those linked to the INDIMO pilot projects (food delivery, ride- and car-sharing, electronic parcel lockers and multimodal route planners), with 'other' as an additional option, to provide the evaluator with the option to evaluate another type of service.

Once this information has been entered in the tool, the actual evaluation starts, in which the evaluator will have to provide answers to the questions linked to each of the evaluation-topics (Table 1). The structure of the questionnaire is threefold, linked to the three themes present within the tool: Universal Design Principles, service features and assistance provided.

4.2 Topics and Questions for Evaluation

As explained in previous sections, input for the tool is based on multiple interviews with each of the stakeholder groups, after which the information was used to develop the UDM, UIL and SCG.

To create a clear structure three themes, based on the collected information, were created, the first theme is the universal design principles, which is based on an adaptation of the standard universal design principles. From the original seven principles, five were kept, the other principles (simple and intuitive, size and space for approach and use) were left out. In the digital world, 'simple and intuitive' was considered to be very different for each stakeholder group and service, also, for a large part, this aspect is covered by the other principles. The 'size and space for approach and use' was not necessary as the physical aspect is not evaluated by the SPET, rather this principle was replaced by digital (and spatial) wayfinding. The last topic that was included in this theme, is the data protection and privacy of the users. Furthermore, one principle was adapted slightly, a cognitive part was added to the low physical effort principle, to better fit with the digital approach.

The second theme 'service features' is an umbrella term for all the topics related to the inclusivity of the services considering pricing, payment methods, information provision, communication and spatial accessibility.

Finally, the third theme 'assistance offered' refers to the topics that provide help for those who have issues using the application or the service, as well as focusing on the iconology used in the application. This topic was identified as a separate topic as one of the key findings of the interviews with users and non-users was that people would need human assistance when using digital mobility services.

This resulted in 18 topics, divided among three main themes as presented in Table 1.

Table 1. Evaluation topics included in the service and policy evaluation tool

Theme	Number of questions	Explanation
Universal Design Principles	34	
Flexibility	6	The design accommodates a wide range of individual preferences and abilities
Equitability	7	The design is useful and marketable to people with diverse abilities
Perceptibility	3	The design communicates necessary information effectively to the user, regardless of ambient conditions or the user's sensory abilities
Tolerance for error	2	The design minimizes hazards and the adverse consequences of accidental or unintended actions
Physical and cognitive effort	1	The design can be used efficiently and comfortably and with a minimum of fatigue
Digital and spatial wayfinding	6	How easily can someone navigate the app and the spatial environment to use the service
Data protection and privacy	9	How privacy and GDPR are taken into account and presented to the user
Service features	27	
Payment	8	The different options for users to pay for their ride, order, trip, subscription, etc
Subscription, reservation & registration	4	The subscription, reservations and registrations options that are available, as well as their usability
Price and affordability	5	The different options of tickets and subscriptions that are available so all people, no matter their financial status can make use of the service
Information	4	The availability and accessibility of information about the service and application
Communication	4	What channels are used for communicating about the service and how effectively info is communicated

(*continued*)

Table 1. (*continued*)

Theme	Number of questions	Explanation
Service area	2	The operational area of a service (geographical and across socio-demographic groups)
Assistance provided	13	
Digital capability	2	The skills needed to access digital tools & use them according to individual needs
Audio assistance	2	Auditory assistance provided within the application to help people (e.g. by telephone)
Autism	2	Specific features of an application to make use easier for people with autism
User feedback	5	How users can provide feedback, how fast operators respond and to what changes feedback lead
Iconology	3	Use of icons has a significant impact on usability, perceptibility,...

Each of these themes, with their appropriate topics form the core structure of the tool. For the evaluation of individual topics, a number of questions were developed for each theme.

4.3 Structure of the Questions

The entire questionnaire is made up of three types of questions, categorized by the way in which they are answered. The first type of questions requires a 'yes or no' answer, the second type are answered in a Likert scale with 3 or 5 potential answers, the last type of questions are the ones that can be answered by selecting suitable option from multiple choices. Once the evaluator has answered all questions the topic and theme-wise scores are calculated.

However, the final score does not only depend on the answers from the evaluator, but it is also influenced by the weights allocated to each of the topics.

4.4　Weighting of the Topics

Each of the topics was described and presented to, and discussed by, 29 experts in the field of digital mobility, policy and inclusivity in order to determine the relative importance of the topics. Experts concluded that all three themes are equally important for an accessible and inclusive application or service, but topics under a certain theme may differ in their relative importance. For allocating weights to each of the topics, a survey was distributed among the experts (operators, developers, researchers and mobility professionals) participating in the INDIMO project. To quantify the relative importance of the topics, the experts were asked the question: 'Relative to the importance of other topics in this theme, how would you score this topic out of 20'. It was decided to use a scale of 1–20 with the consideration and objective of providing experts a scale wide enough to comfortably express the differences in the perceived importance of each topic. Based on the scores provided by experts, the topics within a theme were compared and weights, relative to each other, were allocated to the topics Table 2. The standard deviation of the weights allocated to each topic is also included in Table 2, showing the variance in answers provided by the experts.

Table 2. Allocation of weights to each of the SPET topics

Theme	Topic	Theme weights	Topic weight score on 20	Std Dev	Topic weight
Universal Design Principles	Flexibility	0.33	15	2.95	0.0457
	Equitability		15.64	3.46	0.0476
	Perceptibility		17.5	2.23	0.0533
	Tolerance for error		14.93	2.91	0.0455
	Physical and cognitive effort		15.21	3.38	0.0463
	Digital (and spatial) wayfinding		15.71	2.79	0.0479
	Data protection and privacy		15.43	4.27	0.0470
	Total/Average		**15.63**	3.14	**0.33**
Service features	Payment	0.33	14.43	2.64	0.0540
	Subscription, reservation & registration		13.57	3.54	0.0508
	Price and affordability		15.36	3.04	0.0575

(*continued*)

Table 2. (*continued*)

Theme	Topic	Theme weights	Topic weight score on 20	Std Dev	Topic weight
	Information		16.64	2.58	0.0623
	Communication		14.86	3.4	0.0557
	Service area		14.14	3.83	0.0530
	Total/Average		**14.83**	3.17	**0.33**
Assistance offered	Digital capability	0.33	16.07	3.73	0.0660
	Audio assistance		16.64	3.99	0.0684
	Autism		15.86	3.7	0.0652
	User feedback		16.5	2.87	0.0678
	Iconology		16.07	3.73	0.0660
	Total/Average		**16.45**	**3.50**	**0.33**

Calculating the weights for each topic happens as follows: for the simplicity of calculation and convenience of understanding and presentation, total weight for all themes is considered to be 1. After this the total weight was equally distributed among three themes which are equally important from the perspective of accessibility and inclusivity of an application and service. This way, the weight allocated to each theme is 0.33 (approximately). Then 0.33 was divided among topics according to the ratio of the average score (out of 20), allocated by the experts to each topic to find the topic weights. The actual weight is then calculated by dividing the topic weight score by the sum of all topic weight scores within a theme. The result of this calculation is presented in Table 2 in the column on the right 'Topic weight'.

4.5 Assessment Results and Recommendations

The final score for each theme is calculated based on the answers for each of the questions that topic contains. Table 2 shows the number of questions for each topic, the number of questions does not have an effect on the importance of a theme. Due to the fact that yes/no questions, multiple choice and different Likert scales are used, a transformation is necessary (Table 3). Each of the scores is re-distributed on a 20 point scale. After the re-distribution, the average of all the unweighted score for each question results in the unweighted topic score.

Table 3. Re-distribution answers SPET.

Original answer	Re-distributed answer
Yes – no	20, 0
3 point Likert scale	0, 10, 20
5 point Likert scale	0, 5, 10, 15, 20
Multiple choice	Equal distribution on 0–20 scale (depending on the question: the more/the less, the better)

Once the evaluation of the service is finished, i.e. the evaluator answered all the questions, the average of all weighted topics scores results in the theme performance score. Finally, the average of the three themes results in the overall performance score for the evaluated digital transport service. An example is given in Table 4.

Table 4. Example calculating theme performance score

Topic	Weight score on 20	Topic weight	Score (based on input policy maker) on 20	Weighted performance score on 20	Theme average %
Flexibility	15	0.04524	16	14.48	64.76
Equitability	15.64	0.04717	18	16.98	
Perceptibility	17.5	0.05278	14	14.78	
Tolerance for error	14.93	0.04503	12	10.81	
Physical and cognitive effort	15.21	0.04587	12	11.01	
Digital (and spatial) wayfinding	15.71	0.04738	16	15.16	
Data protection and privacy	15.43	0.04654	8	7.45	

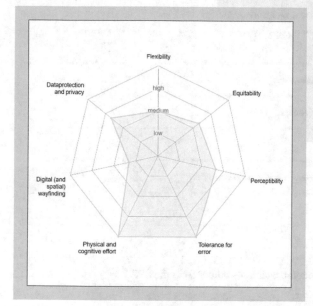

Fig. 3. Example for result spider diagram

After this step all calculations are finished and the results are presented. For each theme a spider diagram is shown (Fig. 3), containing each of the topics included and their performance score in percentage giving the evaluator an easy-to-interpret result. Besides the spider diagram, the evaluator receives the performance scores for each of the three themes, accompanied by a general recommendation, as well as an overall performance score. Three general recommendations are possible, depending on the score of the theme (low, medium or high).

If the evaluator wants a more detailed representation of the results, this is possible as well. For each of the evaluated topics the evaluator receives a score and recommendations that would help to improve the service so that the performance score for a specific topic increases, this of course with the ultimate goal to improve the digital inclusivity and accessibility of the mobility service.

The recommendations are the final result produced by the SPET, providing the evaluator with relevant information on how to make the service more digitally inclusive and accessible. Each of the recommendations (Fig. 4) are specifically linked to the questions in the tool, providing detailed and focused interventions to the evaluator. Depending on the score for each topic different recommendations are presented in three categories: low, med, high, depending on their importance for a more inclusive service.

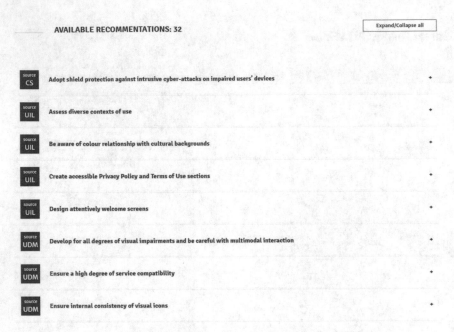

AVAILABLE RECOMMENDATIONS: 32   Expand/Collapse all

source CS   Adopt shield protection against intrusive cyber-attacks on impaired users' devices

source UIL   Assess diverse contexts of use

source UIL   Be aware of colour relationship with cultural backgrounds

source UIL   Create accessible Privacy Policy and Terms of Use sections

source UIL   Design attentively welcome screens

source UDM   Develop for all degrees of visual impairments and be careful with multimodal interaction

source UDM   Ensure a high degree of service compatibility

source UDM   Ensure internal consistency of visual icons

Fig. 4. Recommendations output from the SPET

Recommendations included in the tool are all linked to the aforementioned INDIMO toolbox providing a more elaborate explanation about every recommendation. The recommendations are all organised in the INDIMO toolbox and explained in these documents: (Capaccioli et al. 2020; Ciommo et al. 2020a, b; Hueting et al. 2020).

5   Conclusion and Next Steps

This Service and Policy Evaluation Tool or SPET is the first attempt to develop a tool that provides the opportunity to evaluate the inclusivity and accessibility of a digital mobility service to multiple stakeholder groups. The tool that can be used to evaluate services on multiple topics related to inclusion and accessibility, and we developed a method to quantify these topics. Both the topics and calculating the inclusivity and accessibility score of the services, as well as the recommendations, will provide policy makers and other stakeholders with the framework to efficiently evaluate and score digital mobility services, resulting in the selection of more inclusive and accessible digital mobility services that are allowed to operate.

At the time of writing this paper, the tool was still in its development phase, so no actual testing of the tool has been carried out. The future steps in the development of the SPET are the testing and validation by policy makers and other stakeholders of which the next phase takes places in September 2022. The input from these events will be used for further development of the tool.

A second change that can prove useful is the possibility to adapt the weights distributed between the different topics within a certain range. This way based on the context the evaluator could, to some extent, decide which topics he or she considers more important and thus give a higher weight to that topic. A proposal to do so would be to provide a range, based on the standard deviation from the weighting survey, along which the weight can be changed.

Also, currently, it is only possible to evaluate an entire service, rather than one aspect (application, interfaces and service). In future phases the option will be provided to the evaluator to choose one or more aspect of the service, for which the main motivation is a simpler and faster evaluation for the evaluator.

The most important development that can still be implemented in the tool is the integration of additional services for evaluation. Currently, only the types of services researched in the HORIZON2020 project INDIMO are included, which limits the use of the tool. For future versions, it would be preferable to include other services such as multiple micro-mobility services. At the same time, the SPET can also be adapted to evaluate the services that are not digital, either in the domain of mobility or not. These changes would have a positive effect on the applicability and usability of the tool and should be considered for future research, as well as intensive testing with different services, stakeholders and in a wide variety of cities and regions.

Acknowledgements. The research is a part of the Inclusive Digital Mobility Solutions (INDIMO), a Horizon 2020 project that has received funding from European Union's Horizon 2020 research and innovation programme under grant agreement no. 875533.

References

Aarhaug, J.: Universal design as a way of thinking about mobility. In: Müller, B., Meyer, G. (eds.) Towards User-Centric Transport in Europe. Lecture Notes in Mobility, pp. 75–86. Springer, Cham (2019). https://doi.org/10.1007/978-3-319-99756-8_6

Alkire, S.: The capability approach as a development paradigm? In: Conference on the Capability Approach, Pavia (2003)

Audirac, I.: Accessing transit as universal design. J. Plan. Lit. **23**(1), 4–16 (2008). https://doi.org/10.1177/0885412208318558

Begnum, M., Bue, L.O., Eileen, M.: Towards inclusive service design in the digital society: current practices and future recommendations. In: DS 91: Proceedings of NordDesign 2018, Linköping, Sweden, 14th–17th August 2018. NordDesign 2018. (2018). https://www.designsociety.org/publication/40964/Towards+Inclusive+Service+Design+in+the+Digital+Society%3A+Current+Practices+and+Future+Recommendations

Begnum, M.E.N., Bue, O.L.: Advancing inclusive service design: defining, evaluating and creating universally designed services. In: Rauterberg, M. (ed.) Culture and Computing. Design Thinking and Cultural Computing. Lecture Notes in Computer Science, vol. 12795, pp. 17–35. Springer, Cham (2021). https://doi.org/10.1007/978-3-030-77431-8_2

Beyazit, E.: Evaluating social justice in transport: lessons to be learned from the capability approach. Transp. Rev. **31**(1), 117–134 (2011). https://doi.org/10.1080/01441647.2010.504900

Brown, J.R., Morris, E.A., Taylor, B.D.: Planning for cars in cities: planners, engineers, and freeways in the 20th century. J. Am. Plann. Assoc. **75**(2), 161–177 (2009). https://doi.org/10.1080/01944360802640016

Capaccioli, A., Giorgi, S., Hueting, R., Rondinella, G., Ciommo, F.D.: D2.6 – Guidelines for cybersecurity and personal data protection (public version). Data Protection, 66 (2020)

Chathoth, P., Altinay, L., Harrington, R.J., Okumus, F., Chan, E.S.W.: Co-production versus co-creation: s process based continuum in the hotel service context. Int. J. Hosp. Manag. **32**, 11–20 (2013). https://doi.org/10.1016/j.ijhm.2012.03.009

Ciommo, F.D., Kilstein, A., Rondinella, G.: D2.1 – Universal Design Manual – Version, p. 178 (2020a)

Ciommo, F.D., Rondinella, G., Kilstein, A.: D1.3—Users capabilities and requirements, p. 310 (2020b)

Delaere, H., Basu, S., & Keserü, I. (2020). D1.4 – Barriers to the design, planning, deployment and operation of accessible and inclusive digital personalised mobility and logistics services. 96

Department for Digital, Culture, Media & Sport. Digital Inclusion Evaluation Toolkit. GOV.UK (2017). https://www.gov.uk/government/publications/digital-inclusion-evaluation-toolkit

Durand, A., Zijlstra, T., van Oort, N., Hoogendoorn-Lanser, S., Hoogendoorn, S.: Access denied? Digital inequality in transport services. Transp. Rev. **42**(1), 32–57 (2022). https://doi.org/10.1080/01441647.2021.1923584

Estacio, E.V., Whittle, R., Protheroe, J.: The digital divide: examining socio-demographic factors associated with health literacy, access and use of internet to seek health information. J. Health Psychol. **24**(12), 1668–1675 (2019). https://doi.org/10.1177/1359105317695429

European Commission. Directive (EU) 2019/ of the European Parliament and of the Council of 17 April 2019 on the accessibility requirements for products and services, p. 46 (2019)

European Commission. Directorate General for Mobility and Transport., TIS., Panteia., ITS Leeds., & Armis. (2020). Improving accessibility of persons with reduced mobility by improving digital travel information services: A selection of good practices in Europe. Publications Office. https://data.europa.eu/doi/10.2832/836345

European union agency for fundamental rights. are there legal accessibility standards in place for websites providing public information? European union agency for fundamental rights (2014). https://fra.europa.eu/en/content/are-there-legal-accessibility-standards-place-websites-providing-public-information

Eurostat. Eurostat—Tables, Graphs and Maps Interface (TGM) table (2020). https://ec.europa.eu/eurostat/tgm/refreshTableAction.do?tab=table&plugin=1&pcode=tin00083&language=en

Goodman-Deane, J., Langdon, P., Clarkson, J.: Key influences on the user-centred design process. J. Eng. Des. **21**(2–3), 345–373 (2010). https://doi.org/10.1080/09544820903364912

Gorski, P.: Education equity and the digital divide. AACE J. **13**(1), 43 (2005)

Groth, S.: Multimodal divide: reproduction of transport poverty in smart mobility trends. Transp. Res. Part A: Policy Pract. **125**, 56–71 (2019). https://doi.org/10.1016/j.tra.2019.04.018

Hueting, R., Giorgi, S., Capaccioli, A., Bánfi, M., Soltész, T.D.: D2.3 – Universal Interface Language – Version, p. 212 (2020)

John Clarkson, P., Coleman, R.: History of inclusive design in the UK. Appl. Ergon. **46**, 235–247 (2015). https://doi.org/10.1016/j.apergo.2013.03.002

Kedmi-Shahar, E., Delaere, H., Vanobberghen, W., Di Ciommo, F.: D1.1 –AnalysisFramework of User Needs, Capabilities, Limitations & Constraints of Digital Mobility Services Deliverable D1.1; INDIMO, Nummer D1.1, p. 103 (2020)

Kristensson, P., Matthing, J., Johansson, N.: Key strategies for the successful involvement of customers in the co-creation of new technology-based services. Int. J. Serv. Ind. Manag. **19**, 474–491 (2008). https://doi.org/10.1108/09564230810891914

Kuk, G., Janssen, M.: Assembling infrastructures and business models for service design and innovation. Inf. Syst. J. **23**(5), 445–469 (2013). https://doi.org/10.1111/j.1365-2575.2012.00418.x

Law, C.M., Yi, J.S., Choi, Y.S., Jacko, J.A.: A systematic examination of universal design resources: part 1, heuristic evaluation. Univ. Access Inf. Soc. **7**(1), 31–54 (2008). https://doi.org/10.1007/s10209-007-0100-1

Loos, E., Sourbati, M., Behrendt, F.: The role of mobility digital ecosystems for age-friendly urban public transport: a narrative literature review. Int. J. Environ. Res. Public Health **17**(20), 16 (2020). https://doi.org/10.3390/ijerph17207465

Lucas, K.: Transport and social exclusion: where are we now? Transp. Policy **20**, 105–113 (2012). https://doi.org/10.1016/j.tranpol.2012.01.013

Mace, R.L., Mueller, J.L., Story, M.F.: The universal design file, p. 172 (1998)

Martens, K.: Transport justice: designing fair transportation systems. Routledge, Taylor & Francis Group (2017)

Newman, D.; Top 10 Digital Transformation Trends For 2021. Forbes (2020). https://www.forbes.com/sites/danielnewman/2020/09/21/top-10-digital-transformation-trends-for-2021/

Nussbaum, M.C.: Women and Human Development the Capabilities Approach. Cambridge University Press (2000)

Nussbaum, M.C.: Creating capabilities the human development approach. Orient Blackswan (2011)

Nussbaum, M.C., Sen, A.: World Institute for Development Economics Research (Red.). The Quality of life. Clarendon Press ; Oxford University Press (1993)

Pangbourne, K., Mladenović, M.N., Stead, D., Milakis, D.: Questioning mobility as a service: unanticipated implications for society and governance. Transp. Res. Part A Policy and Pract. **131**, 35–49 (2020). https://doi.org/10.1016/j.tra.2019.09.033

Pappers, J., Keserü, I., Macharis, C.: Co-creation or public participation 2.0? An assessment of co-creation in transport and mobility research. In: Müller, B., Meyer, G. (eds.) Towards User-Centric Transport in Europe 2. Lecture Notes in Mobility, pp. 3–15. Springer, Cham (2020). https://doi.org/10.1007/978-3-030-38028-1_1

Polaine, A., Lovlie, L., Reason, B.: Service design from insight to implementation. Rosenfeld Media (2013). https://public.ebookcentral.proquest.com/choice/publicfullrecord.aspx?p=5198114

Raghavan Sathyan, A., et al.: Digital competence of higher education learners in the context of COVID-19 triggered online learning. Soc. Sci. Humanit. Open **6**, 100320 (2022). https://doi.org/10.1016/j.ssaho.2022.100320

Randal, E., et al.: Fairness in transport policy: a new approach to applying distributive justice theories. Sustainability **12**(23), 10102 (2020). https://doi.org/10.3390/su122310102

Ryan, J., Wretstrand, A., Schmidt, S.M.: Exploring public transport as an element of older persons' mobility: a capability approach perspective. J. Transp. Geogr. **48**, 105–114 (2015). https://doi.org/10.1016/j.jtrangeo.2015.08.016

Schwanen, T., Lucas, K., Akyelken, N., Cisternas Solsona, D., Carrasco, J.-A., Neutens, T.: Rethinking the links between social exclusion and transport disadvantage through the lens of social capital. Transp. Res. Part A Policy Pract. **74**, 123–135 (2015). https://doi.org/10.1016/j.tra.2015.02.012

Scott, M., DeLone, W., Golden, W.: Measuring eGovernment success: a public value approach. Eur. J. Inf. Syst. **25**(3), 187–208 (2016). https://doi.org/10.1057/ejis.2015.11

Sen, A.: Well-Being, agency and freedom: the Dewey lectures 1984. J. Philos. **82**(4), 169 (1985). https://doi.org/10.2307/2026184

Sen, A.: Functionings and Capability. In: Sen, A., Inequality Reexamined 1ste dr, pp. 39–55. Oxford University Press, Oxford (1995) https://doi.org/10.1093/0198289286.003.0004

Sen, A.: Development as Freedom 1 edn. 6th print. Alfred A. Knopf, p. 366 (2001)

Sen, A.: The idea of justice 1. Harvard University Press paperback ed. Belknap Press of Harvard University Press, p. 467 (2009)

Vaidian, I., Azmat, M., Kummer, S.: Impact of Internet of Things on Urban Mobility -, Red, pp. 4–17. HBMSU (2019). https://epub.wu.ac.at/7101/

van Dijk, J.A.G.M.: Digital divide research, achievements and shortcomings. Poetics **34**(4–5), 221–235 (2006). https://doi.org/10.1016/j.poetic.2006.05.004

Vanobberghen, W., et al.: D1.2 – User needs and requirements on a digital transport system, p. 197 (2020)

Vecchio, G., Martens, K.: Accessibility and the capabilities approach: a review of the literature and proposal for conceptual advancements. Transp. Rev. **41**(6), 833–854 (2021). https://doi.org/10.1080/01441647.2021.1931551

Velaga, N.R., Beecroft, M., Nelson, J.D., Corsar, D., Edwards, P.: Transport poverty meets the digital divide: accessibility and connectivity in rural communities. J. Transp. Geogr. **21**, 102–112 (2012). https://doi.org/10.1016/j.jtrangeo.2011.12.005

W3C. Web Content Accessibility Guidelines (WCAG) 2.0 (2008). https://www.w3.org/TR/WCAG20/

W3C. W3C Accessibility Guidelines (WCAG) 3.0 (2021). https://www.w3.org/TR/wcag-3.0/

Integrating Inclusive Digital Mobility Solutions into German Transport Systems

Alexandra Pinto, Joana Leitão, and Benjamin Wilsch[✉]

VDI/VDE Innovation + Technik GmbH, Berlin, Germany
{alexandra.pinto,joana.leitao,benjamin.wilsch}@vdivde-it.de

Abstract. Digital technologies constitute an essential building block for the modernisation of transport systems and their sustainability. Since improvements in usability and accessibility are intrinsic to most digital solutions, their integration in mobility systems may further contribute to an increased inclusiveness. Nonetheless, digital solutions could potentially exclude vulnerable groups by not catering to their specific needs. Inclusiveness can, however, be maximised when individual technologies are established with such goals in mind and are accompanied by the necessary development of the required (hardware) infrastructure.

The widespread introduction of digital applications was triggered by a major German funding programme aiming at enhancing digitalisation in German municipal transport systems to improve air quality. Given the versatility and high adaptability of digital solutions, the substantial progress toward that objective unlocked a high potential for a more inclusive mobility system, so that further improvements in this direction have become low-hanging fruit that can be easily reaped with further projects and funding directed immediately at this objective. The potential of digital solutions to contribute to various societal and thus political objectives, particularly inclusiveness, is identified in this chapter through an analysis of projects that serve as a collection of successful real-world examples.

1 Digital Technologies for Sustainable Municipal Transport Systems

Over the past century, technological progress has created and improved many forms of mobility, causing a constant rise in distance travelled per person (Schäfer 2017). However, the mobility system that emerged does not cater equally to the needs of all (Pooley 2016). Nowadays, Big Data, software solutions and novel (connectivity) hardware have become central digital tools for innovation and progress across all areas of society. In the field of mobility, they have, for example, enabled the combination of different transport modes for a particular trip via multimodal platforms and an increase in efficiency for existing applications.

In order to ensure that the use of these technologies is citizen-oriented and directed at societal goals, political guidance is required. The objectives set by political action are thus instrumental to shaping paths for the development of digital solutions. In particular, funding of innovation and investment projects in the realm of digitalisation can contribute to multiple objectives, including societal ones. As summarised in Fig. 1, digitalisation

I. Keseru and A. Randhahn (Eds.): *Towards User-Centric Transport in Europe 3*, LNMOB, pp. 275–290, 2023.
https://doi.org/10.1007/978-3-031-26155-8_16

can contribute to at least four societal objectives in the field of mobility, wherein the availability of data and cooperative systems presents a prerequisite for the application of many digital solutions, and can, together with advancements in automation and connectivity technologies, serve to enhance the contribution of digitalisation toward these goals. The set of objectives and the prioritisation among them are, of course, subject to change as the societal and political focus shifts. For example, while resilience was never a political focus area until recently, it has gained substantial attraction as a political objective over the past two to three years. The need for this has been particularly emphasised in response to the COVID-19 pandemic, which has highlighted the importance of the ability to adapt to new situations and recover from setbacks quickly. At the same time, integrating aspects of inclusiveness not only into everyday life but also into policies is becoming increasingly more important as diversity is a vital element of an adaptive and innovative society (Llopis 2019).

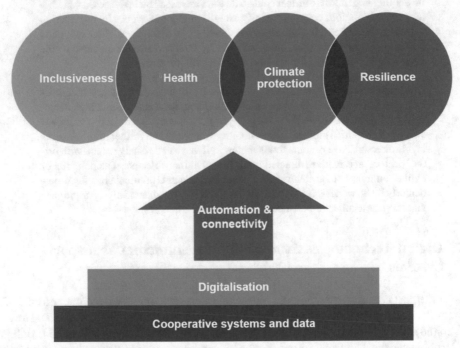

Fig. 1. Building on cooperative systems and the increasing availability of (mobility) data, digitalisation measures can contribute to societal objectives.

In recent years, digital technologies and apps have further become essential tools in addressing the growing urgency of achieving sustainability goals. However, in the course of this digital transformation, inclusiveness and accessibility, major aspects of sustainability, have rarely been amongst the highest priorities. Sustainability does not only pertain to the reduction of the impact of human activities on the environment, but also focuses on social equity and economic feasibility. In a sustainable transformation of society, both its impacts and the social dimension need to be considered, even though they are often not addressed in transition strategies. The potential unemployment and increase in costs are just two examples of consequences that can (and should) be averted when devising balanced, just and systemic approaches that put forward innovative solutions addressing environmental issues and that ensure no one is left behind. To foster social equity in the mobility sector, different aspects of inclusiveness must be taken into account. The transformation of public transportation and reductions of usage costs as well as the improvement and expansion of cycling infrastructures are two examples of inclusive and equitable measures. These minimise community segregation while also reducing urban air pollution levels and contributing to climate change mitigation. However, these do not address the challenges faced by people with visual impairments or those with language barriers. In order to reach the societal and political objective of providing affordable mobility to every member of society and to meet the everyday mobility needs of all, the mobility system must be improved in all aspects, including the modernisation of physical infrastructure.

To address such issues, inclusiveness has been made a general objective in many political agendas across Europe and there is a general growing awareness of its importance. However, it was not until recently that it started to be considered an important aspect to be addressed in sustainability strategies that were once directed primarily, if not exclusively, at achieving particular environmental targets. Nowadays, there is a growing number of different projects that, although not initially conceptualised to address inclusiveness, have succeeded in strengthening the social pillar of sustainability and have indeed contributed to making transport systems more inclusive. The end-effect of these projects can, however, be significantly increased if the aspect of inclusiveness is addressed specifically, and ideally already during the planning phase of the projects, by making use of the right tools.

Digital transformation and digitalisation of systems are seldom oriented towards inclusiveness alone. In the following chapter, we will identify some digitalisation measures that were not specifically developed to support a more inclusive society, but have the potential to do so or are indeed already indirectly contributing. Overall, in this paper, we address in particular the challenges vulnerable groups face in this digital transformation and identify the potential of digital tools to contribute to improved inclusiveness.

1.1 Defining Inclusiveness

Before describing and analysing the main efforts being performed in Germany and in particular their (potential) contribution to improving inclusiveness, it is instructive to recall the objective of inclusiveness first: According to the Cambridge Dictionary (2022), the term 'inclusiveness' is defined as "the quality of including many different types of people and treating them all fairly and equally." Further, Niemann (2019) defines the term of inclusiveness as "the right to be there," meaning that all citizens, including those considered vulnerable users, should have the opportunity to actively participate and to be included as users in all areas of life.

Depending on the context, the definition of vulnerable persons differs widely. The European Parliament for example describes vulnerable adults as persons with severe physical or mental disabilities as well as elderly people who may be limited due to age-related illnesses (EU Parliament 2022). Focusing on vulnerability within the mobility sector, here we follow the definition put forward by the Inclusive Digital Mobility Solutions (INDIMO) project, which identifies vulnerable-to-exclusion individuals as those facing physical, cognitive or socio-economic barriers (INDIMO 2021).

The European Commission published the Strategy for the Rights of Persons with Disabilities 2021–2030 with the goal of enabling people with disabilities to fully take part in society, shaping a fairer and more inclusive Europe (EU Commission 2021). However, many other citizens could also be affected by a lack of accessibility and inclusiveness in various aspects of their everyday life. For instance, with respect to mobility, a parent travelling with their child in a stroller or an elderly person with mobility impairments might not be able to access public transport modes due to physical barriers such as stairs.

1.2 Addressing Inclusiveness

In response to the lack of accessibility, local associations committed to social causes have attempted to address this issue. An example of this is a Berlin-based initiative that allows volunteering participants to record whether or not locations are wheelchair-accessible (Emmett 2021). Collected information on the accessibility of locations is provided to users free of charge while the responsible transport associations or developers are urged to improve accessibility in those locations that were deemed inaccessible.

1.3 Inclusive and Accessible Mobility in Germany

The German government formed in 2021 pledges to "enable sustainable, accessible, innovative mobility catering to the everyday needs of and affordable for all" (SPD, Bündnis 90/Die Grünen, FDP 2021) and points to digital technologies as key elements of innovation for all transport modes.

With respect to accessibility as one component of an inclusive mobility system, societal and political objectives have already been adopted into German law concerning public transport infrastructure. For many modes of transport, physical infrastructure directly determines the level of accessibility and has thus been addressed by policy makers working toward an inclusive mobility system. In Germany, an amendment was made to the German Passenger Transportation Act (in German: "Personenbeförderungsgesetz") on January 1, 2013 in line with the UN Convention on the Rights of Persons with Disabilities (BMJ 2021). According to Scct. 8, local transport plans must take into account the needs of people with reduced mobility or sensory impairments. The aim of which is achieving complete accessibility for the use of local public transport by January 1, 2022. However, with plausible argumentation specifically disclosed in local transportation plans, some exceptions were granted to municipalities that could not meet the deadline stipulated by this act. This was the case for many municipalities and therefore full accessibility has not yet been achieved in many local public transport systems despite the deadline having passed. For instance, 78% of Berlin's 175 subway stations are currently step-free (BVG 2022a). While this seems like a notable number, it also means that more than one in five stations is not accessible for certain passengers, including wheelchair users. In this regard, Hamburg is far ahead of Berlin with 95% of its subway stations being accessible (Hochbahn AG 2022). Meanwhile, among the German cities of over a million inhabitants, Munich leads the implementation rate with 100% of its subway stations being step-free (MVG 2022). Yet, step-free access to subway stations is only one aspect of accessibility and the delayed implementation of the requirements stipulated by the law serves to underline the lack of prioritisation of inclusiveness in the past. In most municipalities, there is, therefore, still a long way to go before full accessibility is reached.

1.4 The Impact of Digitalisation on an Inclusive Mobility Sector

Meanwhile, the digital layer of mobility systems has become increasingly important for the accessibility and inclusiveness of the system and can provide faster-to-implement solutions. Digital tools can also serve to provide alternatives while long-term physical infrastructure measures are completed and can sometimes even make planned measures superfluous. At the same time, it must not be forgotten that the expansion of the digital layer of the mobility system also adds a new potential usage barrier. With the vision of an inclusive society, several European programmes have thus been launched to assure that the majority of the European population acquires basic digital skills by the year 2030 (EU Commission). Despite the progress seen in recent years, the large majority of the global and European population are still not ready for a fully digital society (EU Commission 2022; OECD 2022). At the same time, it is understood that the current technological development and emergence of new digital tools has been crucial for an enhanced inclusiveness of transportation modes. This ongoing digital transformation of systems empowers a variety of sustainable innovations not only related to traffic control, increase of road safety and upgrading of public transport infrastructure, to name a few examples, but could also play a key role in the promotion of inclusive mobility. When supplementing and improving traditional transportation means with innovative digital solutions, it is important to safeguard the inclusiveness of such processes. For example, the development of automated driving technologies, on the one hand, offers the opportunity to be more independent to people with certain disabilities that would otherwise need assistance when traveling. Full autonomous vehicles with guaranteed inclusiveness standards have the potential to be highly advantageous to several vulnerable groups in society, such as elderly people or those with other disabilities, or people that simply do not hold a driving license. On the other hand, digital applications became essential for the improvement of inclusiveness in shared mobility means (e.g. car sharing and on-demand transport), public transportation, and active mobility modes. Examples of this include not only the support of navigation to reach destinations in an effective and secure way, but also those that offer the possibility of different ticketing possibilities (e.g. easy and contactless payment, comprehensible and valid for multi-modal trips, etc.).

Nonetheless, there are many cases in which digital solutions exclude rather than include certain vulnerable groups of society. For instance, apps that are designed without all potential users in mind may not be screen-reader friendly and thus, are not usable for some users with, for example, visual impairments or people that are not digital natives.

2 Digitalisation of Transport Systems in Germany

Funding programmes should foster and encourage activity and investment in areas of high societal relevance and the prioritisation of funded topics should reflect the urgency with which the issues at hand should be addressed. As detailed above, inclusiveness has gained significant momentum with respect to both thematic relevance and urgency over the past years, as shortcomings of current (public) transport systems have become more apparent and as the topic has risen on the political agenda in response to an increased awareness in our society (DIMR 2019). In the following, an overview of major digitalisation measures to improve the efficiency and reduce the environmental impact of mobility systems funded by the German government in recent years is presented. The focus is placed on key areas in which inclusiveness is addressed by these measures, namely: data collection, barrier-free digital platforms and apps, acoustic and visual passenger information as well as traffic management. The overview provides a basis for the subsequent identification of synergies between the different objectives of digitalisation measures as well as for the derivation of recommendations for a systematic approach that maximises contributions to as many objectives as possible, even if one of these is a clear priority for a particular programme.

2.1 Digital Technologies for Health (Clean Air Policies) and Climate Protection

In 2017, the digital layer of mobility systems was identified by the German government as a point of action for a short-term improvement of air quality. Air pollution episodes, namely in urban areas, are a main cause for premature mortalities. Consequently, the European Air Quality Directive on ambient air quality and cleaner air for Europe sets standards for several air pollutants such as nitrogen dioxide (NO_2) and particulate matter (PM). To comply with these thresholds, several measures were implemented in the past decade with the goal of reducing air pollutant emissions from different sources. In this context, and aligned with the overall European trend, the immediate action programme "Sofortprogramm Saubere Luft 2017–2020" (in English: Immediate Action Programme Clean Air) was launched in 2017 by the German government with the goal of improving air quality in the 90 German municipalities that had, at the time, persistent exceedances of, in particular, NO_2 standard levels (above the European Air Quality Standard of $40 \, \mu g/m^3$, as yearly average value). To alleviate air pollution, the programme consisted of three core areas:

- Electrification (740 million euros),
- *Digitalisation (650 million euros),* and
- Hardware retrofitting for public transport diesel buses (107 million euros).

The implementation of the immediate action programme was divided between different ministries and funding programmes, but action plans for improved air quality ("Green City Plans") were required to ensure that a systemic approach was adopted in each city and to maximise the effectiveness of the measures with respect to a short-term reduction of emissions and their effects on human health and the environment. This nationwide German initiative endorsed a broad variety of projects, with the great majority oriented towards electrification of public transport fleets as well as a general modernisation of the mobility sector by making use of innovative digital tools to, for example, promote cycling in cities, by creating faster, safer and easier transportation means. For an effective transition to a sustainable mobility sector, it is crucial that strategy plans contemplate not only the fleet of private vehicles but also focus on the development of alternative mobility modes such as increased use of public transportation or active mobility, i.e. cycling and walking.

The funding budget of 650 million euros in the digitalisation branch of the immediate action programme was administered by the Federal Ministry for Digital and Transport (BMDV) and implemented in the programme Digitalisation of Municipal Transport Systems (in German: "Digitalisierung kommunaler Verkehrssysteme"; DkV) focusing on the promotion of sustainable mobility concepts, with a main orientation towards both the digitalisation of the mobility sector and also strengthening modal shifts in urban environments. With a base funding rate of 50%, recipients of funds from the programme had to match the investment by the German government, so that the overall investment for the digitalisation of transport systems was well above one billion Euro.

The "Green City Plans" concept provides the basis for municipalities' funding applications under the DkV programme. These plans set out a strategic vision for the transformation of urban mobility by making use of integrative approaches that would not only provide solutions for improving urban air quality, but would also shape the overall transportation sector. In doing so, municipal projects make use of digital tools to achieve their long-term goal of becoming more sustainable. Several cities conceived their "Green City Plans" according to the eight principles for Sustainable Urban Mobility Planning (SUMP), a concept defined by the European Commission as part of the Urban Mobility Package. These principles include the development of all transport modes in an integrated manner and, importantly for the context of this article, the involvement of all citizens and stakeholders (Observatory 2021). The ensuing digital transformation has the potential to completely transform the urban environment and improve the life of citizens.

Sustainable mobility plans include many different strategies and creating an attractive network of public transportation is an essential measure to captivate passengers and persuade citizens to reduce usage of private vehicles. A great share of the projects financed within the DkV programme have profited from digital progress and resulted in not only the refurbishment and upgrade of public transport systems themselves, but also of the linked infrastructure. Transformation and modernisation were not only needed in physical infrastructure but especially in the creation of innovative tools that nowadays support and facilitate the life of passengers and cyclists. Indeed, one other important category of implemented measures are those promoting active mobility modes, such as cycling and walking. By turning cycling into a safer, easier and possibly even the fastest transport mode, citizens are more motivated to change their behaviour and use bikes as a means of transportation. In the context of physical infrastructure modernisations, many cities and Hamburg in particular are planning ahead and considering a future of automated driving in their transportation planning. This mainly includes modernisations of traffic light systems, which will enable vehicle-to-everything (V2X) applications (Hamburg 2022). These measures will be especially beneficial for certain vulnerable groups who currently lack independent mobility, but will gain autonomy once autonomous vehicles can safely drive on the road.

Digital tools can also advance the electrification of vehicle fleets, a fast growing trend in recent years and one of the first sustainable solutions considered when designing policy plans to address urban air pollution. Technology developments and widespread digitalisation in industry have been crucial in accelerating the recent transformation of the mobility sector. Digital tools are not only important to manage charging infrastructure but also, for example, assist in the management of electric bus fleets or e-scooters deployed in cities. For example, one of the DkV projects aims to create a virtual net of decentralised charging infrastructure for inner-city public transport by cabs and other (shared) electric transport modes in Dortmund (Dortmund 2022). Multiple cities, including Bonn, Munich and Hamburg, also took advantage of the funding programme to optimise the charging system for their electric bus fleets, in order to improve the availability of the vehicles and to thereby further strengthen the public transport system and its resilience. Meanwhile, e-scooters play an increasing role in first and last mile transport and thus expand the reach of the public network and encourage citizens to switch from their private motorised vehicles to public transportation. When considering the urban logistics and all it entails in current times, apps facilitate not only the management by the delivery companies themselves, but also offer the possibility for recipients to control and check the location of their package and delivery time.

Apps, as one example, be it on a computer or smartphone, are a common ground and key technology in all of these aspects. Such tools nowadays are essential to ensure the smooth functioning of the mobility sector. Yet, the digital transformation of the mobility sector does not rely only on this format and there are many other changes happening at different fronts that are transforming the sector. However, these same tools can also become barriers and create everyday challenges for many citizens when they are poorly designed and not inclusive to all potential users. Here we explore how this modernisation process and all the deployed tools can improve and promote inclusiveness in a digital society based on projects within the DkV funding programme.

3   Potential for Improving Inclusiveness in the Mobility Sector

As previously mentioned, inclusiveness has not generally been accounted for from early stages in the large majority of sustainable mobility oriented policies or conceived projects. Nevertheless, the aspect of accessibility has been addressed by a number of DkV projects. There are currently examples of thirteen projects in eight German cities aiming to directly contribute to the goal of increasing inclusiveness (Fig. 2). The number of such projects is likely to go up as proposals submitted in response to the latest DkV call (May 2022) are, among other criteria, evaluated by their contribution to other goals of the German federal government, such as inclusiveness and resilience in particular.

A common aspect covered in most projects that are currently being implemented is acoustic passenger information, both at stations through speakers and in screen-reader friendly apps. This is, of course, especially beneficial for people with visual impairments. Another point often addressed is step-free access to public transport vehicles, which is particularly crucial for wheelchair users. As a result of the implementation of these projects, public transportation systems have been positively transformed and made more accessible for many citizens. However, accessibility is not the only factor defining inclusiveness. While the introduction of such services and their corresponding apps is a valuable measure to make public transport more attractive, providers seem to often disregard the importance of the accessibility of the apps themselves.

Fig. 2. Map with selected cities that received funding as part of the DkV funding programme of the German Federal Ministry for Digital and Transport (BMDV) for the projects discussed in this chapter.

Still, the digital transformation of the mobility sector holds the potential of improvement in many areas, serving not only transport and transit means but also users of those, i.e., the passengers. This section presents a variety of examples of projects that by addressing digital transformation also promote inclusiveness in the mobility sector.

Here, the focus is on projects that were part of the abovementioned German DkV initiative (from the German name: "Digitalisierung kommunaler Verkehrssysteme") and the classification follows the four main categories of actions that were considered and deployed within this funding programme, (see Fig. 3).

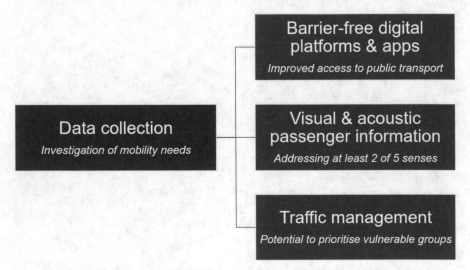

Fig. 3. The four main areas of inclusive digital transformation of the mobility sector within the DkV programme.

Data Collection In order to meet the needs of all citizens, these needs must first be known. Increasing levels of detail concerning the diverse needs and expectations allow for the provision of mobility services tailored to those needs. Since digitalisation and data collection are co-dependent, the initial step of all digitalisation efforts for the mobility sector in all German cities and regions is always the investigation of mobility behaviours. Besides the specific measures with a clearly identifiable contribution to inclusiveness detailed above, the most significant contribution of (funded) digital mobility projects in the long-term will most probably stem from a precise knowledge of (aggregate) mobility needs. All cities that devised a "Green City Plan" for a sustainable mobility system have replaced or are currently replacing manual traffic census with digital and automatic procedures that allow for a manifold increase in temporal coverage and granularity of transport usage patterns (passenger count for public transport or vehicle identification and counting for road systems). Funding from the DkV programme was used by many municipalities to install city-wide sensor networks to accomplish these tasks. This data should prove helpful in conceiving and implementing future measures. For the introduction of new modes and offers, user needs, transport options and the optimisation of city traffic need to be matched in the interest of sustainability goals. With respect to inclusiveness, it must be ensured that data for minority groups is also collected and included in planning steps, since a focus on the needs of the majority would once again leave many individuals behind.

Barrier-Free Digital Platforms and Apps On the basis of the collected data, accessible, comprehensive, transparent and non-discriminatory mobility platforms can be developed. Apps available on smartphones have a multitude of advantages for the use of public transport. The recent digital transformation of the sector and the deployment of many apps have contributed to an improvement of accessibility to public transportation means, by for example offering the information of accessibility of the stations and transport modes. Easier access to public transportation promotes its usage and leads to the increase of the number of passengers. Still, users of vulnerable groups, be it people with disabilities, foreigners that cannot understand the language in which the app is provided, or citizens with low digital skills, are often not considering in the development process of these apps.

Moreover, these platforms should include all mobility providers and all mobility modes. The city of Krefeld in the West of Germany in particular provides a mobility platform which offers ride and travel assistance through an accessible, voice-controlled user interface. In addition, a Bluetooth-based guidance system ensures barrier-free access to buses and trains. Combined with technical equipment in vehicles and buildings, this platform enables direct interaction with vehicles and location-based services. Overall, Krefeld's mobility platform offers a barrier-free assistance system that allows for easy navigation and orientation for both the visually impaired and those unfamiliar with the area.

A multimodality app that harmonises the ticketing and travel information across cities and regions served by different public transportation companies can be helpful to all passengers, and even more to vulnerable users such as elderly citizens. Still, it is important that the app itself is simplified so that users with basic digital skills do not face challenges. The city of Augsburg, for example, is developing a mobility app that combines public transport, car-sharing, and bike-sharing services and thus enables multimodal route planning and simplified billing. In a further implementation stage, the billing function will be supplemented by a "check-in/check-out" function. By combining multiple services in one app, barriers to using public transportation are broken down and the attractiveness of the mobility service is thus increased.

In fact, many apps nowadays provide a large amount of information on stations and stops, namely how accessible they are to people with particular disabilities. In this setting, on-demand transport systems that use barrier-free and inclusive vehicles can also serve to enhance the accessibility and reach of public transport networks, such as those offered in at least two German cities: Berlin (BerlKönig) and Hamburg (Ioki and VHH). The Digitaler Pendlerbus (Digital Commuter Bus, in English), also known as the BerlKönig BC, also functioned as a barrier-free on-demand ridesharing service in Berlin at a price comparable to that of a standard public transport ticket (BVG 2022b). This service was meant to be an extension to the existing public transport system, however, certain vulnerable user groups also benefitted from the service by simply indicating that they require an accessible vehicle during the booking process in the app. At the same time, this type of on-demand service is also beneficial for users living in rural areas who gain better public transport links. This effect could be increased, especially with the introduction of autonomous on-demand shuttles that do not require a human driver to be present anymore. Additionally, Berlin's public transport provider is currently developing the Alternative Barrier-Free Transportation (in German: "Alternative Barrierefreie Beförderung"; ABB) which can be ordered via phone or app (BVG 2022c). This service offers a mobility guarantee for passengers who are dependent on accessibility on the subway, local and regional trains. It is especially useful at stations that are not yet barrier-free or where a lift is currently out of order or being modernised.

Other examples of mobility apps include those for mobile ticketing and cashless payment for public transportation, Park and Ride (P + R) as well as ridesharing. While these apps can be very convenient for able-bodied people and can increase their willingness to use more climate-friendly modes of transportation, they are oftentimes not as helpful for people with disabilities if implemented with limited or no accessibility. Meanwhile, vulnerable persons such as those with disabilities may be the ones who could potentially benefit from such services the most as they could achieve greater autonomy by using digital mobility solutions (Disabled Living 2019). For example, instead of attempting to navigate within a station and to find a physical ticket machine, people with visual impairments could simply purchase a ticket using their smartphone and voice recognition or a screen reader. However, a study by the Research Institute for Disabled Consumers found that 26% of people with disabilities face difficulties accessing or managing smartphone apps (RIDC 2020). Common barriers include complicated navigation links, missing or inaccurate alt text or gifs that could even cause epileptic seizures (Accessibility 2022).

Finally, delivery apps can provide an immense level of independence, particularly for those members of society who are unable to go shopping on their own. An example of this is the INDIMO pilot project conducted in Madrid which assesses a previously existent delivery platform for goods in regard to user experience and user needs (INDIMO 2022d). Especially elderly people face barriers in accessing solutions such as this one due to low digital skills. Based on the results, the platform should then be optimised in order to be more inclusive and accessible for vulnerable persons.

Visual and Acoustic Passenger Information Amongst those DkV projects that set out inclusiveness as one of their key objectives, the majority addresses the introduction of visual and acoustic passenger information. Berlin's public transport provider has incorporated a number of graphic elements in digital passenger information displays, which, for example, indicate whether subway cars are accessible for wheelchair users. Furthermore, passenger information is provided not only visually via text and icons on displays but also acoustically and in multiple languages. Thus, a multitude of public transport users, including foreigners, can perceive this information. In response, the local association for the blind expressed their support of the implementation of these components. Similar projects are being conducted in the German district cities of Limburg, where all indicators are being equipped with pushbuttons and connected to existing guidance systems for the blind, and Gross-Gerau, where text-to-speech software is used for announcements at stations.

Traffic Management Last but not least, initiatives in the field of traffic management mainly focus on road transport sectors and most projects address traffic lights systems, and implementing different forecasting systems. Still, these, and namely the data collected from devices installed in these projects, can prove useful for other applications that promote a wider inclusiveness beyond improving the flow of traffic on roads. Traffic light system data and high-precision intersection maps are made available to other service providers. Subsequently, this can be used as basis for the development of apps and other platforms that serve non-motorised mobility users. The city of Hamburg, for example, is testing an app that informs drivers and cyclists whether the next traffic light will be red or green when they reach it. In the case of cars, emissions can thus be lowered as unnecessary speeding before red lights would be reduced. In terms of cycling, such apps can make the use of bicycles more appealing as cyclists know in advance whether to speed up a little or to take their time in order to reduce the waiting time traffic lights (Reutlingen, 2022). With this data already being collected in many cities, it would be fairly simple to take a step further and create apps that could, for example, help people with visual impairments cross the road safely. To implement this, the information of whether a traffic light is red or green could be provided in an accessible app. Additionally, intelligent traffic lights could prioritise traffic streams and adapt the length of green periods for people who may require some extra time to cross the road, as is being explored in the INDIMO pilot project in Antwerp (INDIMO 2022a).

The listed examples indicate that there is a clear, although far from inevitable, intrinsic link between increasing digitalisation and improved inclusiveness. Although the project examples from the DkV programme were designed to maximise contribution to the very urgent short-term objective of improving air quality – which they did successfully, considering that the number of German cities in which European air quality levels were exceeded has dropped significantly from 90 to below 10 between 2016 and 2021 – they also contributed to other societal goals, notably inclusiveness and thereby improving sustainability overall. The examples therefore demonstrate that digital solutions are a fundamental enabler for inclusive mobility systems, while also underlining that the extent of the contribution depends strongly on the timing and scope of the integration of inclusiveness in project planning.

4 Outlook

The digital transformation of society and particular sectors is in full flow. This unlocks the potential for the improvement of the lives of many citizens and could make Europe a frontrunner in reaching sustainability goals. Mobility in particular is a central enabler for an improved quality of life (Lee and Sener 2016) and the improvement of the reach and resilience of transport systems directly affects its inclusiveness. The digital layer of transport and mobility systems opens up many new applications and provides the necessary levels of versatility and redundancy to address diverse user needs. However, digitalisation does not yet equally affect everyone, with some vulnerable user groups seeing their needs neglected.

Ongoing measures have already led to visible improvements regarding accessibility in the mobility sector. While the presented projects cannot be considered best-practice examples for implementing inclusiveness as this was never their main objective, they do, however, demonstrate that digitalisation efforts bring with them a potential for enhanced inclusiveness that can be unlocked with comparably low effort once the necessary level of connectivity and the digital layer of hardware and software components have been established. Fuelling overall investments above one billion Euro in the digitalisation of municipal transport systems, the DkV funding programme discussed above has laid this groundwork of digitalisation in over 70 major cities in Germany. As presented, in some projects and cities, inclusiveness was already a secondary objective of the measures directed primarily at improving air quality, specifically projects focused on data collection, barrier-free platforms and apps, visual and acoustic passenger information and traffic management. Perhaps more importantly, each project has increased the level of digitalisation in the cities and thus provides the basis for future measures to improve inclusiveness of mobility systems, as the topic gains societal and thus political relevance. The step from digital mobility solutions to inclusive mobility is, of course, not a given and the former will only be achieved, if inclusiveness is indeed prioritised on the agendas of local, regional or (supra-)national governments. Furthermore, the earlier inclusiveness is considered in the planning of mobility projects, the larger their contribution will be in this regard.

The path to a fully inclusive digital transformation is still long and will not always be smooth. Nonetheless, increasing the inclusiveness of transport systems will come with many benefits, especially for certain vulnerable groups whom it could enable to reach a new level of independence. The added value goes beyond the technology in regards to social aspects, which may not always be acknowledged. While here the focus was inclusiveness of digital tools in the mobility sector alone, these notions can be observed in other sectors and many aspects related to everyday life of citizens. Lessons learned in the mobility sector could thus help improve other services, including everything that is nowadays done through apps or online in general, be it shopping, bank transactions or simply booking an appointment with a doctor. Ultimately, the overall quality and comfort of life of vulnerable persons could be drastically improved by acknowledging their needs. In light of this, it could prove helpful to create platforms for knowledge transfer across stakeholders and key actors, enabling the identification of efficient measures and implementation of policies for a transition to an inclusive digital society.

By acknowledging that a digital transformation has overarching impacts, solutions to address current lack of inclusiveness in whatever sector can be better designed. Holistic and systematic approaches have proven to be beneficial when addressing such complex topics as inclusiveness, which involves a variety of sectors and actors. Although there are already success stories in recent developments and innovation efforts, there is still room for improvement. Policies need to be designed in such a way to not target single objectives, but instead consider broader impacts, bringing benefits on various fronts. A set of policy guidelines is needed to assist in the implementation of sustainability measures and assure that the social component of sustainability strategies is considered. In view of this, funding programmes in particular should incorporate aspects of inclusiveness. By doing so, the importance of social sustainability is emphasised and its implementation ensured. A good example for this is the investment in improvement and modernisation of public transport systems that did not only target the reduction of air pollution and overall impacts on the environment, but made use of that opportunity to also promote social equity. Expansion of urban cycling infrastructures, even if mainly focused on the improvement of the health of residents and environmental benefits due to the subsequent modal shift, also aids in connecting communities and improving accessibility. Bringing city and transport planning together from the start is an approach aligned with the concept of smart cities, based on holistic development concepts aiming at making cities more efficient, technologically advanced, more environmentally friendly and socially inclusive. Another example is the concept utilised in designing Sustainable Urban Mobility Plans (SUMP), which are strategic and integrated approaches dealing with urban mobility (POLIS and Rupprecht Consult 2021). These plans aim at improving accessibility and promoting a shift towards sustainable mobility by advocating for seven principles of resilience, including inclusiveness. All of these initiatives are valuable and legitimate to improve inclusiveness in the mobility sector. Still, it is important not to forget to confer with end-users directly, especially those belonging to vulnerable groups, when developing digital mobility solutions to achieve inclusive and accessible tools, so that everyone's needs can be considered and nobody is left behind.

References

Accessibility (2022) Barriers to Independent Living: Digital Accessibility. https://www.accessibi lity.com/blog/barriers-to-independent-living-digital-accessibility. Accessed 19 May 2022

BMJ (2021) § 8 PBefG - Einzelnorm. https://www.gesetze-im-internet.de/pbefg/__8.html. Accessed 19 May 2022

BVG (2022a) Mit der BVG barrierefrei durch Berlin. https://www.bvg.de/de/service-und-kontakt/ barrierefrei-unterwegs. Accessed 19 May 2022

BVG (2022b) Berlkoenig | Die Idee. https://www.berlkoenig.de/die-idee. Accessed 25 May 2022

BVG (2022c) Kommt wie gerufen - BVG Unternehmen. https://unternehmen.bvg.de/pressemittei lung/kommt-wie-gerufen/. Accessed 25 May 2022

Cambridge Dictionary (2022) inclusiveness. https://dictionary.cambridge.org/dictionary/english/ inclusiveness. Accessed 25 May 2022

DIMR (2019) Analyse: Wer Inklusion will, sucht Wege. https://www.institut-fuer-mensch enrechte.de/fileadmin/user_upload/Publikationen/ANALYSE/Wer_Inklusion_will_sucht_W ege_Zehn_Jahre_UN_BRK_in_Deutschland.pdf. Accessed 31 May 2022

Disabled Living (2019) The Benefits of Technology for Disabled People - Disabled Living. https://www.disabledliving.co.uk/blog/benefits-of-technology-for-disabled-people/. Accessed 19 May 2022

Dortmund, T. (05. 31 2022). VIZIT - Virtual integration of decentralized charging infrastructure in cab stands. https://cni.etit.tu-dortmund.de/research/projects/vizit-1/

Emmett (2021) Offene Daten für mehr Barrierefreiheit beim Bus- und Bahnfahren. https://emm ett.io/article/offene-daten-fuer-mehr-barrierefreiheit-beim-bus-und-bahnfahren. Accessed 19 May 2022

EU Commission 2030 Policy Programme, "Path to the Digital Decade - Digital Education Plan - European Pillar of Social Right Action Plan - Digital Compass". In:

EU Commission Union of Equality: Strategy for the Rights of Persons with Disabilities 2021–2030 (2021) https://ec.europa.eu/social/main.jsp?catId=738&langId=en&pubId= 8376&furtherPubs=yes. Accessed 5 Oct 2022

EU Commission Shaping Europe's digital future: Human Capital and Digital Skills in the Digital Economy and Society Index (2022). https://digital-strategy.ec.europa.eu/en/policies/desi-human-capital. Accessed 19 May 2022

EU Parliament Carriage details | Legislative Train Schedule (2022). https://www.europarl.eur opa.eu/legislative-train/theme-area-of-justice-and-fundamental-rights/file-protection-of-vul nerable-adults. Accessed 19 May 2022

Hamburg, C.O. Test track for automated and connected driving in Hamburg (2022). https://tavf. hamburg/en/

Hochbahn AG Barrierefreier Ausbau | Hamburger Hochbahn AG (2022). https://www.hochbahn. de/de/projekte/barrierefreier-ausbau. Accessed 19 May 2022

INDIMO D1.1 – Analysis Framework of User Needs, Capabilities, Limitations & Constraints of Digital Mobility Services (2021)

INDIMO Antwerp - INDIMO Project (2022a). https://www.indimoproject.eu/pilot-project/ant werp/. Accessed 19 May 2022

INDIMO (2022d) Madrid - INDIMO Project. https://www.indimoproject.eu/pilot-project/mad rid/. Accessed 3 August 2022

Lee, R.J., Sener, I.N.: Transportation planning and quality of life: Where do they intersect? Transp. Policy (Oxf) S, **48**, 146–155 (2016)

Llopis, G.: Without Inclusion, Humankind Is Becoming Less Resilient. Forbes (2019)

MVG Barrierefreiheit (2022). https://www.mvg.de/ueber/engagement/barrierefreiheit.html. Accessed 19 May 2022

Niemann, J: 1. Regionalkonferenz Mobilität in einem inklusiven Sozialraum Dokumentation (2019)

Observatory, E. –T. The SUMP Concept. Von (2021). https://www.eltis.org/mobility-plans/sump-conceptabgerufen

OECD Chapter 4. Digital uptake, usage and skills | OECD Digital Economy Outlook 2020 | OECD iLibrary (2022). https://www.oecd-ilibrary.org/sites/def83a04-en/index.html?itemId=/content/component/def83a04-en#section-81. Accessed 19 May 2022

POLIS and Rupprecht Consult Topic Guide: Planning for more resilient and robust urban mobility online version (2021)

Pooley, C.: Mobility, transport and social inclusion: lessons from history. SI, **4**, 100–109 (2016)

Reutlingen, C.O.: Synchronized traffic lights for cyclists (2022) https://www.reutlingen.de/gruene welle

RIDC Research shows quarter of disabled people unable to use key apps | RiDC. (2020). https://www.ridc.org.uk/news/research-shows-quarter-disabled-people-unable-use-key-apps. Accessed 19 May 2022

Schäfer, A.W.: Long-term trends in domestic US passenger travel: the past 110 years and the next 90. Transportation **44**(2), 293–310 (2017). https://doi.org/10.1007/s11116-015-9638-6

SPD, Bündnis 90/Die Grünen, FDP (2021) Mehr Fortschritt wagen – Bündnis für Freiheit, Gerechtigkeit und Nachhaltigkeit

Author Index

A
Aarhaug, Jørgen 157
Abraham, Michael 142
Acosta, Esau 127
Alčiauskaitė, Laura 173
Alegre Valls, Lluís 42
Andersen, Kristina 173

B
Basu, Samyajit 254
Bulanowski, Kathryn 111

C
Capaccioli, Andrea 194
Ciprés, Carolina 3

D
de la Cruz, Teresa 3
Delaere, Hannes 254
Delespaul, S. 235
Devis, Juanita 127
Di Ciommo, Floridea 93, 127
Dubbert, Jörg 22

F
Foldesi, Erzsébet 22

G
Giorgi, Sabina 194
Goodman-Deane, J. A.-L. 235
Graf, Antonia 215

H
Hansel, Julia 215
Hatzakis, Tally 173
Hiemstra-van Mastrigt, Suzanne 74

H
Hoeke, L. 235
Hueting, Rebecca 194

J
Jaenike, Miguel 127

K
Keseru, Imre 59, 254
König, Alexandra 173

L
Lamoza, Thais 127
Launo, Carolina 173
Lazzarini, Boris 235
Leitão, Joana 22, 275
Lima, Sandra 111

M
Marlier, Evelien 111
Martinez, Lluis 59
Moertl, Peter 22

N
Nesterova, Nina 235
Nys, Arne 127

P
Pinto, Alexandra 22, 275

R
Randhahn, Annette 22
Rondinella, Gianni 93, 127
Royo, Beatriz 3

S
Sampimon, Max 74
Sanyer Matias, Xavier 42
Schröder, Carolin 142

© VDI/VDE Innovation + Technik GmbH 2023
I. Keseru and A. Randhahn (Eds.): *Towards User-Centric Transport in Europe 3*, LNMOB, pp. 291–292, 2023.
https://doi.org/10.1007/978-3-031-26155-8

Printed in the United States
by Baker & Taylor Publisher Services